# ENERGYBOOK #1

## Natural Sources & Backyard Applications

David Roditi

edited by John Prenis

Windworks

**Dedication**

This book is respectfully dedicated to all the backyard experimenters and basement tinkerers who have shown that alternative energy research need not be the exclusive domain of "experts."

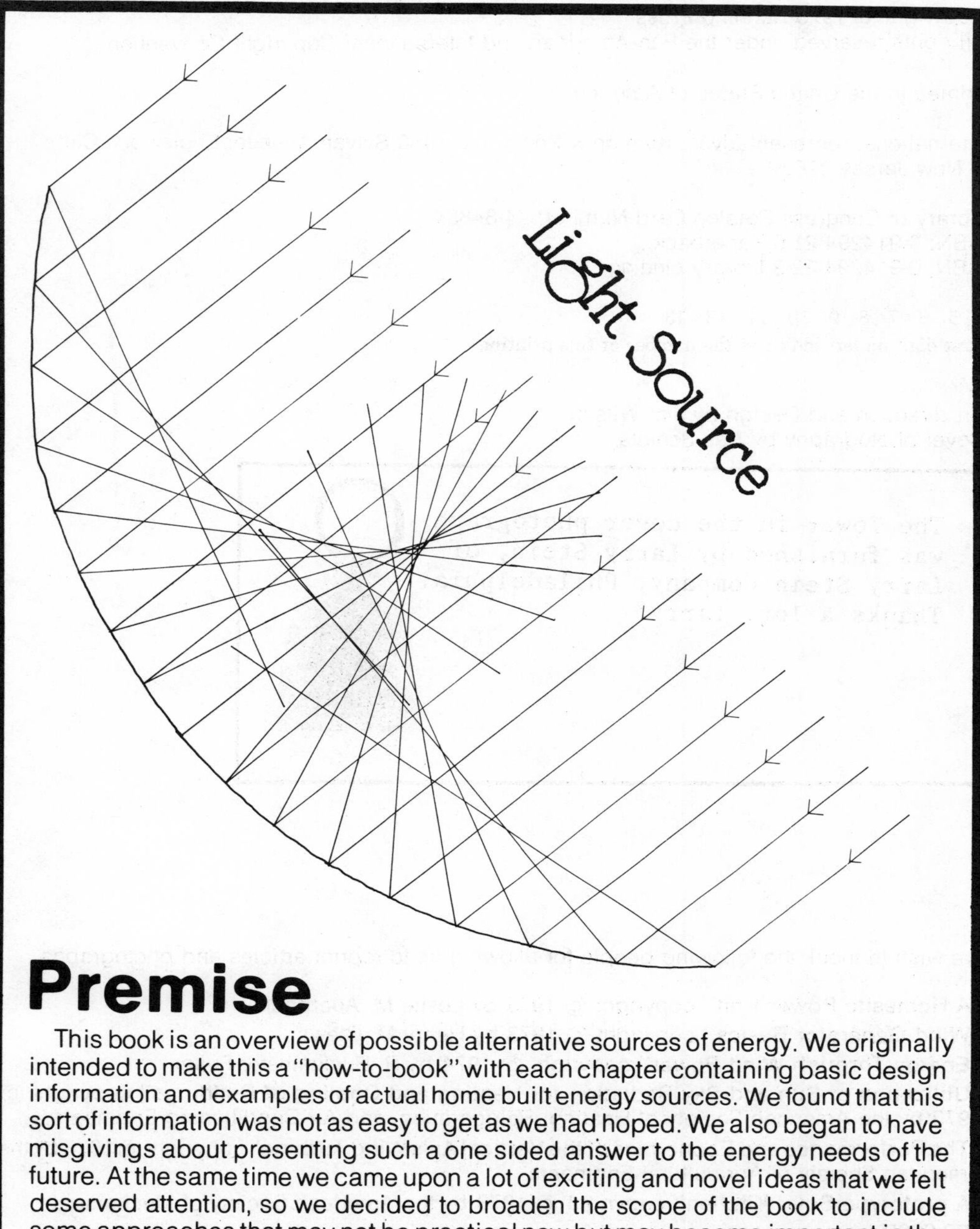


# Premise

This book is an overview of possible alternative sources of energy. We originally intended to make this a "how-to-book" with each chapter containing basic design information and examples of actual home built energy sources. We found that this sort of information was not as easy to get as we had hoped. We also began to have misgivings about presenting such a one sided answer to the energy needs of the future. At the same time we came upon a lot of exciting and novel ideas that we felt deserved attention, so we decided to broaden the scope of the book to include some approaches that may not be practical now but may become important in the future. We hope that this book will appeal both to the interested layman and to the backyard experimenter.

Printed in the United States of America

International representatives: Kaimon & Polon, Inc., 456 Sylvan Avenue, Englewood Cliffs, New Jersey 07632

Library of Congress Catalog Card Number: 74-84854
ISBN: 0-914294-21-0 Paperback
ISBN: 0-914294-22-9 Library binding

5 6 7 8 9 10 11 12 13
First digit on left indicates the number of this printing

Art direction and Design by Jim Wilson
Cover photography by Kas Schlots

The Tower in the cover photograph was furnished by Larry Stein, of Larry Stein Company, Philadelphia. Thanks a lot, Larry.

We wish to thank the following people for allowing us to reprint articles and photographs

"**A Homesite Power Unit**" copyright © 1973 by Leslie M. Auerbach
"**Wind Generator Basics**" copyright © 1973 by Henry M. Clews
"**Energy Through Wind Power**" copyright © 1974 by R. Buckminster Fuller
"**Utilization of Sun and Sky Radiation for Heating and Cooling of Buildings**" copyright © 1973 by the American Society of Heating, Refrigeration, and Air-Conditioning Engineers
"**The Performance and Economics of the Vertical Axis Wind Turbine**" copyright © 1973 by the American Society of Agricultural Engineers
"**A prototype Solar Kitchen**" copyright © 1973 by the American Society of Mechanical Engineers

This book may be ordered directly from the publisher.
Please include 25¢ postage.
**Try your bookstore first.**

Running Press, 38 South Nineteenth Street,
Philadelphia, Pennsylvania 19103

# TABLE OF CONTENTS



Uncredited items in the above by John Prenis.

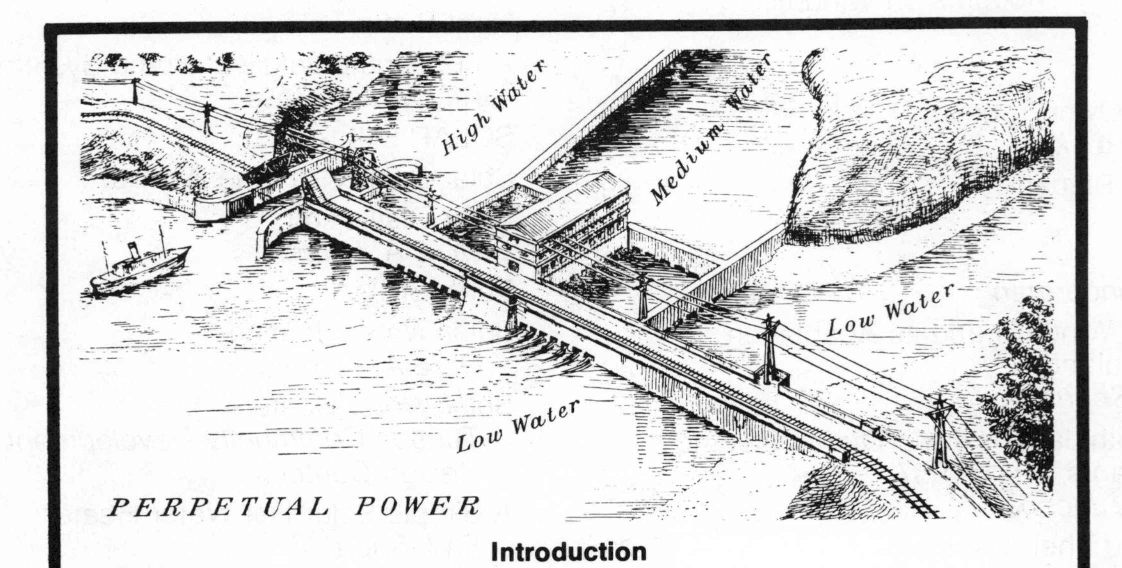


*PERPETUAL POWER*

## Introduction

The development of civilization can be traced through the story of man's discovery and use of new energy sources. The discovery of fire, the domestication of animals, the development of agriculture, the invention of the steam engine, and the use of coal and oil are all important milestones. Each step has brought about a dramatic change in man's way of life. We are now on the threshold of a shift from oil, gas, and coal to new sources of energy. Our industrial civilization was built on abundant supplies of fossil fuels and it has come to depend on them. The first signs of the coming changeover have inspired confusion, uncertainty, and predictions of doom and disaster. The problem is worldwide and affects all of industrial civilization. We've never faced anything like it before. Yet the solution is not beyond our reach. We have the knowledge and the ability to draw energy from the sun, the wind, the tides, and the internal heat of the earth. But we must not become complacent. Heedless use will quickly exhaust our energy resources. During the next few decades we will have to learn how to use energy more wisely, while at the same time new sources of energy are developed.

Many people have become interested in alternative sources of energy because they would like to become more self sufficient. While no one can build his own atomic power plant, a do-it-yourselfer can build his own solar collector, or windmill, or methane generator. But beware of the "build your own windmill from junk parts and tell the power company to jump" fantasy. Major energy projects require knowledge and experience which can best be gained while working with less ambitious projects. We hope readers will try their own experiments, but as a means of self education and discovery. Someone working with alternative energy sources can hardly help but develop a new sensitivity to the flow of energy in the world around him, and an awareness that energy is something to treasure, not waste. And there is nothing like the thrill of capturing energy that did not come out of a pipe or a wall socket.

# ENERGY

Energy is usually defined as the ability to do work. That leaves us with the problem of defining work. To a physicist, work is force times distance. A force of one pound acting through a distance of one foot has done one **foot-pound** of work. Work can be thought of as energy in action. It is convenient to measure energy and work in the same units, so we can say that one foot-pound of energy is needed to do one foot-pound of work. Usually we think of work in a more general sense. We say we have to do work in order to set something into movement, change its motion, heat it up, or change its shape. Work is also needed to compress a gas, stretch a spring, or to drive the chemical processes that charge a storage battery.

There are many different forms of energy. There is mechanical energy, the energy possessed by moving objects. There is electrical energy, which drives electrons through wires. There is chemical energy, released when unstable chemical compounds change to more stable forms. There is heat energy, the random motion of molecules in matter. All of these forms of energy can be measured in the same units, and each can be converted from one form to another.

Energy is a comparatively recent concept. Only within the past century have two important basic facts been established. The first was that despite appearances, energy is never really "used up". Instead, it is transformed into some other form. It took many very careful experiments to establish this, but it was always found that the "missing" energy that was lost through incomplete combustion, friction, air resistance, etc. could be accounted for as noise, vibration, stray heat, and so forth. By keeping careful track of energy losses, it was also discovered that energy was never created either. No matter what you do, the amount of energy you wind up with is never more than what you start with. In fact it is precisely the same, every time. This discovery put an end to the old dream of the perpetual motion machine that would do work without requiring any energy input. This "conservation of energy" has survived every test, and as extended in the **conservation of mass-energy**, it is one of the cornerstones of modern physics.

The demonstration of the conservation of energy did not kill the dream of unlimited free energy. For instance, there is energy all around us in the form of the heat contained in every object by virtue of the random motion of its molecules. It has always been a dream of inventors to harness this heat energy in a process something like the following: you'd take a boiler full of water, extract the heat from part of it, and use it to boil the rest. Half the boiler's contents would turn to ice, the rest to steam, which could then be used to drive an engine. The ice would be dumped out, the boiler refilled, and the process started all over again. You would have an unlimited supply of energy and all you'd need would be a lake of water to turn to ice. This doesn't violate the principle of conservation of energy either, since you wouldn't be creating or destroying any energy but just using what was there in the first place.

Unfortunately this happy idea is not practical for another reason. Heat will only flow from warm objects to cool ones, just as water will not flow uphill by itself. It has been found that spontaneous energy processes will go in only one direction. That direction is the one which results in energy becoming less concentrated, more disorganized. The ultimate tendency of energy is to spread out as thinly and evenly as possible. Concentrated energy is useful for doing work, but diffuse energy is useless. It's the difference between a mass of water held behind a dam and a flooded plain. It's true that energy processes can be **made** to go in the other direction, just as water can be pumped uphill, but the energy required to do this becomes dispersed and scattered in its turn and the net result is less useful energy. It's a case of one step forward, two steps back.

Your refrigerator is a good example. It makes heat flow in the "wrong" direction, from the cold inside to the warm outside. This difference in temperature represents energy, but it is established at the cost of dissipating energy elsewhere, as you discover when the time comes to pay your electric bill.

A practical result of this is that no energy transformation can be 100% efficient. After a long series of energy conversions only a fraction of the original energy is available in the desired form. The rest has been transformed into non-useful forms of energy, and will probably end up as stray heat. Heat is the most disorganized form of energy at all, since it consists only of random molecular motion.

To sum up, there are two fundamental limits on what we can do with energy. First, you can never have more or less than you started with. Second, every energy process scatters some energy, which cannot be regained. No exceptions to these two principles have ever been discovered. They have achieved the status of scientific laws—the first and second laws of thermodynamics.

## Power

To say that a job takes a given amount of energy tells us nothing about how quickly the job can be done. A stringbean who takes a quarter of an hour to stagger to the top of a 20 foot staircase with a 100 pound weight is obviously in a different class than a stevedore who whisks the same weight up in less than a minute, even though both have done 2000 foot-pounds worth of work. We say that the stevedore is more powerful than a stringbean. Power is the **rate** at which work is done and is measured in joules/sec, ft-lbs/sec, or horsepower. The joule/sec is usually called the **watt**, after James Watt, who developed the horsepower as a means of comparing the output of steam engines with that of more familiar power sources. One horsepower is the power needed to raise a weight of 33,000 pounds 1 foot in 1 minute.

(Actually one horsepower is more than the average horse can put out on a continuous basis. Watt's horses were large dray horses that had a chance to rest between trials. In case you're interested, you can put out about 1/10 hp continuously if you're in good shape; much more for a short period of time.)

Many people talk loosely about power and energy as though they were the same. Strictly speaking, this is not correct. What the power company sells you is **energy**. That is why your electric meter reads in kilowatt-hours. The kilowatt, a unit of energy/time, is multiplied by a period of time to turn it back into energy. If we needed to, we could measure energy in hp-second, or watt-minutes.

The generating capacity of power plants is measured in kilowatts or megawatts to show how much energy a plant is capable of putting out in a given time.

You might find it fun to measure your own short term power capacity by running up a flight of stairs while a friend times you with a stopwatch. Your weight times the height of the staircase divided by the time it took you gives you the power output in ft-lbs./sec. Divide this by 550 to give you your power output in horsepower.

## Units of Energy and Power

In the early days of technology everyone invented a different unit to measure what he was doing. These arbitrarily defined units were followed by scientifically defined ones. In addition, there are metric and English system units for the measurement of each quantity. The result is that today we have our choice of five different units in which to measure energy, and two units in which to measure power. What follows in a brief description of the background of each unit.

We have already discussed the foot-pound, which is the work done when a force of one pound acts through a distance of one foot. The corresponding metric unit is the newton-meter. This is the work done by a force of one newton acting through a distance of one meter. The newton-meter is better known as the joule, named after James Prescott Joule, who demonstrated the equivalence between work and heat. Another metric unit of energy is the erg, which is one ten-millionth of a joule. The erg is too small for all but certain scientific measurements, so it is seldom seen.

A good deal of energy is used to heat water. Here the British thermal unit (BTU) is helpful. One BTU is the amount of heat energy required to raise the temperature of one pound of water 1°F. The corresponding metric unit is the calorie. One calorie is the amount of heat energy needed to raise the temperature of one gram of water 1°C. More useful is the kilocalorie, or large calorie, which equals one thousand calories. To avoid confusion it should be mentioned that the kilocalorie is the "calorie" used to measure the energy contained in foods and that dieters count. Careless usage has worn the "kilo" off the front of the word. This should be kept in mind when comparing the heat energy available in food with other sources, or otherwise the figures will seem one thousand times too small.

The situation is simpler when talking about power. There are only two principal units—the horsepower and the watt. One horsepower is 33,000 foot-pounds per minute, or 550 foot-pounds per second. A watt is one joule per second. We also hear of kilowatts and megawatts, which are one thousand watts and one million watts, respectively. Since power is actually energy per unit time, it is also possible to measure it in such units as BTU's per second or joules per day, but this is not often done.

The fact that power is energy per unit time also gives us still another way to measure energy. Any unit of power multiplied by a unit of time becomes a unit of energy. The kilowatt-hour is the familiar example. It is equivalent to one thousand watts times the number of seconds in an hour or 3,600,000 joules. Energy could also be measured in kilowatt-weeks or horsepower-hours, but there is seldom any good reason for doing so.

All these different units cause difficulty when it is necessary to compare figures. It is hoped that the following conversion tables will be of some help.

### WORK, ENERGY, HEAT

| | joule | ft-lb | kilo-cal | BTU | kw-hr |
|---|---|---|---|---|---|
| 1 joule | 1 | 0.7376 | $0.2389 \times 10^{-3}$ | $0.9481 \times 10^{-3}$ | $0.2778 \times 10^{-6}$ |
| 1 ft-lb | 1.356 | 1 | $0.3239 \times 10^{-3}$ | $1.285 \times 10^{-3}$ | $0.3766 \times 10^{-6}$ |
| 1 kilo-cal | 4186 | 3087 | 1 | 3.968 | $1.163 \times 10^{-3}$ |
| 1 BTU | 1055 | 777.9 | 0.2520 | 1 | $0.2930 \times 10^{-3}$ |
| 1 kw-hr | $3600 \times 10^3$ | $2655 \times 10^3$ | $860.1 \times 10^3$ | 3413 | 1 |

### POWER

| | hp | ft-lb/sec | kw |
|---|---|---|---|
| 1 hp | 1 | 550 | 0.7457 |
| 1 ft-lb/sec | $1.818 \times 10^{-3}$ | 1 | $1.356 \times 10^{-3}$ |
| 1 kw | 1.341 | 737.6 | 1 |

| Metric Prefix | Common usage | Scientific notation |
|---|---|---|
| micro | 1 millionth | $10^{-6}$ |
| milli | 1 thousandth | $10^{-3}$ |
| kilo | 1 thousand | $10^3$ |
| mega | 1 million | $10^6$ |

### Some Energy Equivalents

1 ton bituminous coal = $26.2 \times 10^6$ BTU
1 barrel crude oil = $5.6 \times 10^6$ BTU
1 barrel residual oil = $6.29 \times 10^6$ BTU
1 gallon gasoline = $125 \times 10^3$ BTU
1 gallon No. 2 fuel oil = $139 \times 10^3$
1 cubic foot natural gas = 1031 BTU
1 kw-hr electricity = 3413 BTU

### MORE ENERGY EQUIVALENTS

| | |
|---|---|
| 1 Ton TNT | $39.8 \times 10^6$ BTU |
| 1 megaton nuclear bomb | $39.8 \times 10^{12}$ BTU |
| 1 gram of matter completely converted to energy | $85.3 \times 10^9$ BTU |

# Energy Through Wind Power

By R. Buckminster Fuller

The inventions that impound solar energy through Sun-reflecting or lensing devices are fascinating to the imagination but relatively insignificant in potential. In fact, they work only a few hours daily when the Sun is at a favorable angle.

Even if half of Arizona were turned into a direct-sunlight-energy-converting mechanism, the production would be negligible in comparison with windpower sources.

Wind power is in a class by itself as the greatest terrestrial medium for harvesting, harnessing and conserving solar energy. The water and air waves circulating around our planet are unsurpassed energy accumulators whose captured energy may be used to generate electrical, pneumatic and hydraulic power systems.

Windmills produce power from the Sun-generated differentials of heat, which are the source of all wind, with far greater efficiency than do attempts to focus and store direct solar radiation. But the most comprehensive consideration regarding wind power is not technological. Rather it is an appreciation that wind power is by far the most efficient way to recapture solar power.

In the first place, three-quarters of the Earth is covered with water, and the remaining quarter is land area consisting largely of desert, ice and mountains. Only about 10 per cent of our planet's area has terrain suitable to cultivation, in which vegetation can impound the Sun's radiation by photosynthesis.

Among the solar-energy impounders in vegetation, none can match corn's performance. Corn converts and stores as recoverable energy 25 per cent of the received ultraviolet radiation, whereas wheat and rice average only 18 to 20 per cent. From these stores of solar energy humans can produce commercial alcohol, or they can leave the energy to the production of fossil fuels in the Earth's crust, which requires millennia.

But one-half of the vegetation-producing area of the Earth's surface is always in the shadow, or night, side, which reduces to 5 per cent the working area of the Earth's surface on which vegetation impounds the Sun's energy. Though theoretically 5 per cent of the area can impound energy at any one time, only an average of one per cent of the Sun's energy is actually being converted because of local weather conditions and infrared and other energy-radiation interferences.

The area of the surface of a sphere, is exactly four times the area of the sphere's great-circle disk, as produced by a plane cutting through the center of the sphere. The surface of a hemisphere is, then, twice the area of the sphere's great-circle plane. When we look at the "full" moon, we are looking at a surface twice the area of the seemingly flat, circular disk in the sky.

All of the Earth's energy comes from the stars, but primarily from the star Sun, as radiation or as inter-astro-gravitational pull. Twenty-four hours a day the Sun is drenching the outside of the hemisphere of the cloud-islanded atmosphere's 100-million-square-mile surface area, which is twice that of the disk of the Earth's profile.

This gives us one billion cubic miles on the sunny side and one billion cubic miles on the shadow side. The atmospheric mass is kinetically accelerated in the hemisphere constantly saturated by the Sun, while simultaneously the atmospheric kinetics in the night atmosphere are decelerated.

All around the Earth, yesterday's Sun impoundment pertubates the atmosphere by thermal columns rising from the oceans and lands. The shadow side consists of one billion cubic miles of contracting atmosphere, while the one billion cubic miles on the sunny side is sum-totally expanding. This rotation of the Earth brings about a myriad of high-low atmospheric differentials and world-around semi-vacuumized drafts, which produce the terrestrial turbulence we speak of as the weather.

The combined two billion cubic miles of continual atmospheric kinetics converts the solar energy into wind power. Wind power is Sun power at its greatest, by better than 99 to 1.

All biological life on planet Earth is regenerated by star energy, and overwhelmingly by the Sun's radiation. The Sun radiates omnidirectionally 92 million miles away from Earth, with only two-billionths of its total radiation impinging upon Earth. The radiation arrives at a rate of two calories of energy per each square centimeter of Earth's Sun-side hemispherical surface per each minute of time. About half of that is reflected back omnidirectionally to the universe. The other half, i.e., one calorie per minute per square centimeter, is impounded by our planet's biosphere in ways making them available to human use, granted humanity's permitted comprehensive, ecologically considerate, employment of its inventive capabilities and spontaneously cooperative potentials.

No matter how dubious one may be of such logical realizations of our potentials, the fact remains that our net receipt and impoundment of cosmic energy amounts to 120 trillion horsepower, which can also be stated as 90 trillion kilowatts, which, with 8,760 hours a year amounts to 786 quadrillion kilowatt-hours a year. This is $786 \times 10^{15}$ kilowatt-hours, almost two hundred thousandfold the world's present $5 \times 10^{12}$ kilowatt-hour production of electric energy.

If all humanity enjoyed 1973's "highest" living standards—that of the United States—each human on Earth would consume 200,000 ($2 \times 10^5$) kilocalories a day.

Assuming five billion ($5 \times 10^9$) humans by A.D. 2000, each consuming $2 \times 10^5$ kilocalories daily, we will need $1 \times 10^{15}$ kilocalories a day, while our actual daily terrestrial income of cosmic energy is $2 \times 10^{18}$ kilocalories. Our planet's usable daily energy income is therefore $2 \times 10^3$, or two thousandfold our daily requirements of A.D. 2000.

I have been pursuing the subject of wind power in varying degrees of intensity since 1927 when I included windmills, air compressors, liquid oxygen liquefaction equipment, and air turbines in the design of the first Dymaxion house.

Concurrent with the advent of rural electrification—more than a third of a century ago—I saw that windmills were going out just as modern aerodynamic research was coming in. To take advantage of this potential scientific harvest, I have for the past four years been pursuing the development of windmill-generated electricity. One of the strategies has been to convert pure Sun-distilled water electrolytically into hydrogen directly for power purposes; or the hydrogen and oxygen could be reassociated to produce electric current at an overall 85 percent efficiency. A complementary strategy is the combination of improved, variable pitch, windmill propeller blades with the aerodynamics of jet technology in which the wind-flow patterns embracing whole buildings are captured and funneled into low-pressure focusing, Pitot-tube, cowlings. We are also developing a new octahedral windmill mast which is transportable, powerful, economical, and swiftly erectable; as well as a new low-cost method of mechanical linkage from the mill to the generator.

In the course of our experiments, my associates and I found that the Greek-island type windmills with self-furling sails are also very efficient.

Present experiments show that flywheels—as energy accumulators—can be employed efficiently in connection with variable winds to drive generators.

All the winds around Earth together with all the force they can use to produce the 150 million square miles of ocean waves; and to bend, twirl, twist and sometimes uproot the world's trees, bushes, grasses, dust storms; and to form and scud around the Earth the two billion cubic miles of clouds and their many violent storms; as well as the billions of tons of water raised hourly into the sky to rain back upon Earth to maintain the vegetation; these winds and their many side-effect tasks altogether constitute a 100-mile thick, 200 billion-cubic-mile spherical man t le which is indeed a Sun-energy accumulator or Sun-energy storage battery, whose power capacity is adequate to accommodate and eternally regenerate all of humanity's needs and pleasures, with a safety factor coefficient of 10,000 to 1.

As the U.S. Navy reckons it, one minute of one hurricane releases more energy than that of the combined atomic bomb arsenals of the United States and Russia. From the viewpoint of design science, it is simply a matter of coping with the calm, zephr, gale, or hurricane variabilities of wind power.

Great corporations have not as yet ventured into this field because wind energy has not seemed to be monopolizable over a pipe or a wire. Enterprise can be rewarded, however, in greater magnitude than ever before, by producing and renting world-around wind-harnessing apparatus—following the models of the computer, telephone, car rental, and hoteling-service industries.

Hydrogen, harvested in the manner I have described, can be used immediately to operate all the world's piston- or turbine-driven engines now driven by gasified petroleum products. Recent studies by the National Science Foundation afford statistical confirmation of my statement that wind power can take care of all our energy needs, and that this can be accomplished in short order. The National Science Foundation's development strategy is directed at producing large offshore, ship- or tower-mounted, windmill batteries to supply large cities. In contradistinction, my windmill development work is aimed to supply individual consumer families.

Wind power permits humanity to participate in cosmic economics and evolutionary accommodation without in any way depleting or offending the great ecological regeneration of life on Earth.

*– R. Buckminster Fuller*

---

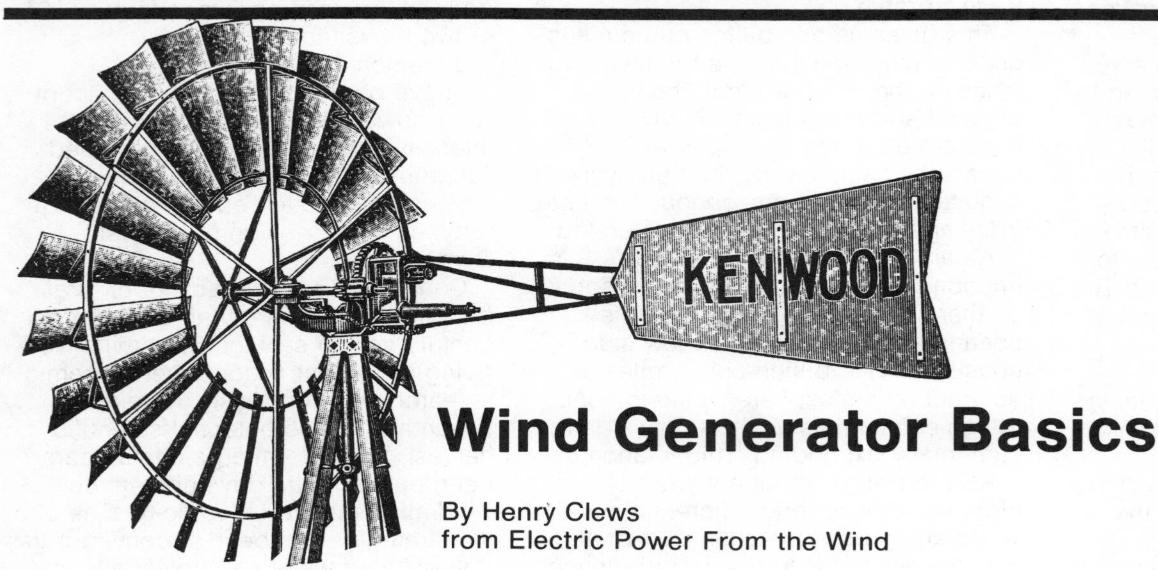


# Wind Generator Basics

By Henry Clews
from Electric Power From the Wind

Virtually all electricity is produced by rotating electric generators which produce electricity by rotating magnets in front of each other. The Power Company uses huge generators turned by steam turbines, (or in the rare case of hydroelectric power, by water turbines). The steam to turn the turbines is produced by boiling water over a coal fire (hiss, boo!). New atomic energy power plants function in the same way except that the heat to produce the steam comes from radioactive fission instead of coal. In an automobile, the generator is turned through a V-belt by the gasoline engine. Similarly, in a small portable power plant, the generator is turned by a gasoline, diesel, or L.P. gas engine. Thus all forms of useable electricity comes from some type of a rotating generator which is driven by an external power source. The wind generator is no exception. A wind driven generator consists of a rotating generator turned by a propeller which in turn is pushed around by the force of the wind upon it. The propeller can be thought of as a wind engine using wind as its only fuel.

Now, the amount of electricity that can be generated by a wind generator is dependent on four things: the amount of wind blowing on it, the diameter of the propeller, the size of the generator, and the efficiency of the whole system. Here are some specific examples to show you how this works. First consider an 8-foot diameter propeller with well designed blades having an efficiency of, say, 70% and a generator capable of delivering 1000 watts. In a 5 mph breeze you might get 10 Watts of power from it; at 10 mph about 75 Watts; at 15 mph, 260 Watts; and at 20 mph, 610 Watts. As you can see, the more wind the more power. But it is not a simple relationship. The actual power available from the wind is proportional to the **cube of the wind speed**, in other words if you double the wind speed you will get eight times as much power.

Now let's consider a propeller with a 16-foot diameter and a similar efficiency to the first one. At 5 mph wind we might get 40 Watts output; at 10 mph, 300 Watts; at 15 mph, 1040 Watts; and a 20 mph, 2440 Watts if the generator were capable of delivering this much power. As you can see the power output of the 16-foot diameter windmill is about four times that of the 8-foot diameter windmill. This shows that the power is proportional to the **square of the diameter**, or that doubling the size of the propeller will increase the output by a factor of four. And there you have the two basic relationships which are fundamental in the design of any wind driven power plant. A careful study of the table below will further serve to illustrate these relationships.

| Propeller Diameter in feet | Wind Velocity in mph | | | | | |
|---|---|---|---|---|---|---|
| | 5 | 10 | 15 | 20 | 25 | 30 |
| 2 | 0.6 | 5 | 16 | 38 | 73 | 130 |
| 4 | 2 | 19 | 64 | 150 | 300 | 520 |
| 6 | 5 | 42 | 140 | 340 | 660 | 1150 |
| 8 | 10 | 75 | 260 | 610 | 1180 | 2020 |
| 10 | 15 | 120 | 400 | 950 | 1840 | 3180 |
| 12 | 21 | 170 | 540 | 1360 | 2660 | 4600 |
| 14 | 29 | 230 | 735 | 1850 | 3620 | 6250 |
| 16 | 40 | 300 | 1040 | 2440 | 4740 | 8150 |
| 18 | 51 | 375 | 1320 | 3060 | 6000 | 10350 |
| 20 | 60 | 475 | 1600 | 3600 | 7360 | 12760 |
| 22 | 73 | 580 | 1940 | 4350 | 8900 | 15420 |
| 24 | 86 | 685 | 2300 | 5180 | 10650 | 18380 |

TABLE 1 — WINDMILL POWER OUTPUT IN WATTS assuming 70% efficiency

But what about efficiency and generator size? The efficiency (defined as the ratio of the power you **actually get** to the theoretical maximum power you **could** get at a certain wind speed) depends largely on what type of propeller you use. All modern electric wind generating plants use two or three long slender aerodynamically shaped blades resembling an aircraft propeller. These efficient propellers operate at a high **tip speed ratio** which is the ratio of propeller tip speed to wind velocity. The Quirk's propeller, for example, runs at a tip speed ratio of about 6 while for some of the Swiss Elektro units this ratio runs as high as 8. This compares to ratios of 1 to 3 for the slower running multi-blade American water-pumping windmill. But while the latter type is less efficient they do have a much higher starting torque and their steadier speed at low wind velocities makes them more suited for pumping applications.

Ideally, the propeller of a wind machine used for generating electricity should have a cross-section resembling that of an aircraft wing, with a thick rounded leading edge tapering down to a sharp trailing edge. It should be noted, however, that the most efficient airfoils for aircraft propellers, helicopter blades, or fan blades (all designed to **move** air) are not the most efficient airfoils for windmills (which are intended to be **moved by** air). An old airplane propeller, in other words, has neither the proper contour nor angle of attack to satisfactorily extract energy from the wind. If you have the ability and time, and wish to construct a propeller of your own, you will find several good designs to copy in the United Nations publication ***Proceedings, Vol. 7, UN Conference on New Sources of Energy***.

The generator itself forms the vital link between wind power and electrical power. Unfortunately, most generators which would seem to be suitable suffer from the requirement that they need to be driven at high speeds; they are built to be driven by gasoline engines at speeds from 1800 rpm to 5000 rmp. But windmill speed especially in the larger sizes, seldom exceeds 300 rpm. This means that one must either find special low speed generators (which are expensive and cumbersome) or resort to some method of stepping up the speed of the generator using belts, sprockets, or gears. The large commercially available units generally make a compromise here. They use a relatively low speed generator (1000 rpm) and they gear the generator to the propeller through a small transmission at about a 5 to 1 step-up ratio.

The next question is, how to decide what size generator to use with what size propeller. Well, here again some compromises are in order. First, you must decide what windspeed will be required for your generator to put out its full electrical output. If you want full output at low wind speeds you will need a large propeller, whereas if you are satisfied with full output only at high wind velocities, a small propeller will suffice. In general light winds are more common than strong winds. Statistical studies of wind data (from Dayton, Ohio) show that each month there is a well-defined group of wind velocities which predominate. These are called the **prevalent** winds. There is also a well-defined group which contains the bulk of the energy each month, called, appropriately enough, **energy winds**. The first group, consisting of 5 to 15 mph winds, blows 5 out of 7 days on the average, while the energy winds of 10 to 25 mph blow only 2 out of 7 days. It might seem, at first glance that you should design for maximum output at, say, 15 mph to take advantage of all those prevalent winds, but this would require a very large propeller for the power produced, and all the power from winds higher than 15 mph would be thrown away. As an example, consider a 2000 Watt generator which is to yield its full output at 15 mph. From Table 1 we can deduce (assuming a 70% efficient system) that to get 2000 Watts at 15 mph we will need something over a 22 foot diameter propeller. Now this will be large and expensive to build, and difficult to control in high winds. Besides, look at all that power you are throwing away at higher wind speeds. If you really are going to build a 22 foot diameter rotor, then you might as well install a bigger generator on it and get some of that power at higher wind speeds, right?

Well, in practice this is what is done. Most working wind generators are designed to put out full power in wind speeds of about 25 mph, and in so doing they do sacrifice some performance at low wind speeds. Usually they deliver almost no output at wind speeds below 6 or 8 mph, but this is really not a serious drawback because there is so little energy available from these light winds anyway. Now you can begin to understand just what is meant by a "2000 Watt" wind generator. As you can see it is hard to compare the rated output of a wind generator to that of a conventional generating plant with the same rating. In the case of the wind generator, the power rating merely tells you what the **maximum output** of the generator will be at a certain wind speed—and you must know what this wind speed is if you want to calculate how much power you will actually get from a certain windplant under varying wind conditions.

This brings us to the final problem in choosing a suitable wind electric system for your homestead. The question is, how much total electric energy will a certain size system produce over a period of time in your particular location? This is the main concern of anyone attempting to determine the feasibility of a wind generator in their area—and it is also the most difficult question to answer. Suppose you have sat down and, you have figured out that, to run everything you want to run in your new wind powered homestead, you will need 200 Kilowatt-hours of electricity per month. If all your power is to come from the wind, you will need a system that will provide at least this much per month and a little more besides to allow for the slight inefficiency of the storage batteries.

Well, to actually figure out precisely how much a certain system will deliver in a given location, you must know not only the complete output characteristics of your wind generator at different wind speeds, but also you must have complete windspeed data for the proposed installation site. And by complete, I mean enough data to plot a continuous graph of the wind speed for a year or two. Such a graph would allow you to compute the total energy available from the wind in your particular location. Roughly, the power available would correspond to the area under the curve, but even this is not mathematically correct because of the cubic dependency of power on wind speed. To do it right, would require some pretty sophisticated statistical analysis which we will certainly not venture into here—and actually it all becomes pretty academic since few people have good enough wind data for their location, anyway. But, if you **do** want to know more about this, Putnam's book, **Power from the Wind** will be of some help.

Now, lest you begin to despair, let me give you some idea of how to procede in the absence of all the facts. This might be considered "fudging it", but unless you're planning a very expensive commercial installation, it will certainly get you into the right ball park. First, find out what the **average** yearly winds are in your location. The Weather Bureau records wind speeds hourly at several hundred stations across the country, and if you write them (see source 2) you can get this information including average wind speeds for each month and year at a station near you. Use this as a start, but don't consider it definitive. Winds at your actual site may vary considerably from those at the local weather station, so you will probably want to carry out some tests of your own, especially if you are in a doubtful area, ie. official average winds much under 10 mph. Later we will discuss measuring wind speed and selecting the best site in more detail, but for now let's assume you have satisfied yourself that the average winds at your location are, say, 12 mph. Well, we have prepared a handy-dandy little table based on our limited experience in this field, which will hopefully give you some idea of what you can expect from different size windplants at various average wind speeds. As you will appreciate, many factors enter into this and we have had to make several assumptions. First, these figures are based on typical present production wind generator designs with tip speed ratios on the order of 5 and efficiencies of about 70%. It is also assumed that there is negligible output below wind speeds of 6 mph and that maximum output is reached at 25 mph. This table represents a composite of actual measurements, plus some figures put out by several wind generator manufacturers plus a fair amount of interpolation.

| Nominal Output Rating of Generator in Watts | Average Monthly Wind Speed in mph | | | | | |
|---|---|---|---|---|---|---|
| | 6 | 8 | 10 | 12 | 14 | 16 |
| 50 | 1.5 | 3 | 5 | 7 | 9 | 10 |
| 100 | 3 | 5 | 8 | 11 | 13 | 15 |
| 250 | 6 | 12 | 18 | 24 | 29 | 32 |
| 500 | 12 | 24 | 35 | 46 | 55 | 62 |
| 1000 | 22 | 45 | 65 | 86 | 104 | 120 |
| 2000 | 40 | 80 | 120 | 160 | 200 | 235 |
| 4000 | 75 | 150 | 230 | 310 | 390 | 460 |
| 6000 | 115 | 230 | 350 | 470 | 590 | 710 |
| 8000 | 150 | 300 | 450 | 600 | 750 | 900 |
| 10,000 | 185 | 370 | 550 | 730 | 910 | 1090 |
| 12,000 | 215 | 430 | 650 | 870 | 1090 | 1310 |
| TABLE 2 | AVERAGE MONTHLY OUTPUT IN KILOWATT-HOURS | | | | | |

As I said, this table should only be considered as a rough estimate of what you can expect from wind generated power in different wind areas. Many manufacturers of wind generators refuse to commit themselves to anything as specific as the figures listed in this table because they claim that conditions vary so much, what with the effect of turbulence, temperature, etc., that they would only be sticking their necks out to make any specific predictions of long term energy output. Nevertheless, we feel that this is the one basic statistic that everybody wants to know when they consider installing a wind electric system and so have included this table for your use.

Now we may proceed to use the table to solve the original problem which was, how large a system will you need to get that 200 KW-hrs per month in an area with a 12 mph average wind speed? Checking Table 2 under the 12 mph column we find that a 2000 Watt system would produce only 160 KW-hrs while a 4000 Watt generator would produce 310 KW-hrs. Interpolating between these two values we can estimate that a 3000 Watt unit might produce 230 KW-hrs per month which is just about right allowing for the inefficiencies of batteries, inverters, etc. Of course, when it comes around to buying or building such a system you may be forced by financial considerations, or by what is actually available, to install a larger or smaller system, but at least you'll have some idea what you can expect from it when it's all done.

Before we move on to the next subject, here are a few comments about wind generators in high winds. Perhaps, as we were talking about maximum outputs at 25 mph and such, you were wondering what happens at higher wind speeds. Well, all modern production windplants are designed to function completely automatically in winds up to at least 80 mph or even higher, so rest assured that there is no such thing as a site with **too much** wind. In order to survive all kinds of winds, wind generators employ some method of holding down their speed in heavy winds. The most common method of spilling excess wind, whenever the power from the wind exceeds the power rating of the generator, is a system of weights mounted on the propeller which act centrifugally to change the pitch of the blades, thus reducing the wind force on the propeller. This system, which amounts to a built-in governor, holds the propeller at a constant speed and prevents overspeeding when there is little or no load on the generator—which happens whenever the batteries are fully charged and no power is needed. This is one area where the modern windplant has come a long way in solving a problem that plagued the windchargers of forty years ago. Burned out bulbs and even burned out generators were not uncommon with the old units as the windmill raced out of control in heavy winds.

But even with the modern version, the manufacturers generally recommend that if wind speeds greater than 80 mph are anticipated, as in a hurricane, the propeller should be manually stopped and/or rotated sideways to the wind. Most models have a brake control located at the bottom of the tower for this purpose. Such "furling" of the windmill during a storm greatly reduces the strain of high wind loads on the propeller and on the entire tower structure.

*This is only a part of the booket "Electric Power From the Wind". You can get the complete booklet plus a packet of information about commercial wind generators for $2 from Solar Wind, P.O. Box 7, East Holden, Maine 04429.*

# Some Cautionary Thoughts on Windmills

Windmills have been pumping water, grinding grain, and doing other simple mechanical work since the 12th century. They are now receiving a great deal of attention as a means of easing the energy shortage.

Windmills fall into two main groups: those that do mechanical work, and those that generate electricity. The first type is the kind once found on many farms for pumping water. It has many wide flat blades and turns slowly. It produces high torque, which makes it well suited for many simple chores. The large area it presents to the wind enables it to turn even in light breezes. Because it runs slowly, careful balancing and precise construction is not essential.

The second type of windmill is more recent. Designed to produce electricity, it is more properly called a wind generator. Electric generators must spin at high speed for best efficiency. Special low speed generators can be built, but they are heavy, bulky, and expensive. It is easier to design the blades for high speed, which accounts for the wind generator's distinctive appearance. It usually has two or three long slim blades which closely resemble those of an airplane propellor. It turns at high speed, which means that it needs careful balancing and precision construction or it will vibrate badly and shake itself to bits. Although its high speed simplifies connection to a generator, it still usually requires extra gearing between the blades and the generator.

The excitement of the wind enthusiast usually turns to dismay when he looks at the prices of commercial units. These are ruggedly built pieces of precision machinery, designed to run year after year in all sorts of weather with minimal maintenance—and their prices reflect the fact. The ambitious amateur is often inspired to try building his own wind generator, but this is not as simple as it seems. To begin with, the high speed of a wind generator makes its construction more critical. A wind generator involves both a mechanical system and an electrical system and both of them have to work properly. (The fact that the system is more complex than a simple water pumping windmill means that it is both harder to get working and more likely to break down.) There are some tricky engineering problems to solve, and trying to make do with scrounged or salvaged parts doesn't help any. If the system is built around an alternator from a junked car, its power output will be limited to less than a kilowatt, which is not much return for the time and work invested. I don't mean to say that a practical home built wind generator is impossible. Those who are mechanically inclined and good with tools will see the problems as challenges and have fun conquering them. For those just starting out, though, I would recommend a simpler project such as a Savonius rotor. A low speed windmill can do simple mechanical jobs and provide valuable experience.

# Windworks

*Windworks is a research-oriented collective concerned with the development of structural systems that allow man to live in greater harmony with his world and with the transfer of information that allows the greatest learning and experiencing on the part of those giving and receiving.*

*Our work on renewable energy sources began under the direction and sponsorship of Bucky Fuller with the development of a paper honeycomb technique of blade construction which allows individuals to construct low-cost, efficient airfoils. Our current work includes developmental programs, information distribution, prototype design and construction, and aeronautical, mechanical, and structural engineering of wind installations.*

*The following is a review of the windmills Windworks has built and the experience gained from them.*

The design, construction and operation of wind driven generators is a complex process even in a production situation. Home building shares the difficulties but offers a number of unique advantages. By investing your time, the cost of a wind powered generating plant can be reduced by a factor of 3 to 5. The most valuable aspect of the process is the learning involved.

Decentralization, self-sufficiency, energy consumption patterns, and the complexity of on-site power generation become very real.

The power outputs given are power available from the blades. Usable power depends on the drive train and generating or driven equipment efficiencies. In home building, where substitutions are often made on the basis of availability of components, blade power outputs are the basis for determining usable power. These outputs can be calculated for a given type of wind machine, diameter, and wind speed using the power equation in the appendices.

We are interested in encouraging the further development and use of wind energy and will be happy to help those working in this direction.

Windworks Box 329 Route 3
Munkwonago, Wisconsin 53149

## Venturi (shrouded) Wind Generator

*Description*
***shroud***
inlet 6 ft. diameter
outlet 7 ft. 2 in. diameter
length 8 ft. 3 in.
The shroud was built using typical aircraft construction; plywood bulkheads and stringers, skinned with muslin, fiberglass and resin.
***blades***
2 bladed, 5 ft. diameter
NACA 4415 profile
mounted 30% back from inlet, paper honeycomb core with fiberglass — resin skin
***tower***
30 ft. 2 frequency tetrahedron, spruce pole construction, free standing
***other***
power transmission via V-belt blades mounted on motorcycle hub alternator mounted in shroud
***materials cost***
$500

### Design Objectives

Our primary interest was in investigating the practicality of the shrouded wind generator. It was shown theoretically in the 1950's that this type of design could increase the power output by a factor of 2 over an unshrouded wind generator of the same diameter. This is a result of the increased flow velocity through the venturi, the reduction of blade tip losses, and the control of the slip stream expansion down stream of the rotor. This was also the first trial for the paper honeycomb core/fiberglass skin technique of blade construction.

### Experience

The shroud proved overly sensitive to sudden changes in wind direction and required damping in the main bearing at the top of the tower. Power outputs were approximately 1.5 times those calculated for a free-standing mill of the same diameter. Because of the time and cost involved in constructing the shroud, this design could be most sucessfully applied in a production situation. It offers advantages over a free-standing mill when space is limited or safety from the spinning blades is a prime consideration.

### References

1. A Preliminary Report on the Design and Performance of Ducted Windmills, G.M. Lilley and W.J. Rainbird, 1957, Electrical Research Association Technical Report C-T 119.
2. **Wind Energy**, Hans Meyer, Dome Book II, 1971

## 10 Foot Diameter Solid Foil

*Description*
***blades***
3 bladed, 10 ft. diameter Wortmann FX-60-126 profile paper honeycomb-fiberglass construction mounted downwind of tower
***tower***
10 ft., 2x4 wood construction, guyed
***other***
chain and sprocket power transmission 9:1 step-up with jack shaft 12 volt 35 amp automotive alternator
***materials cost***
$180

### Design Objectives

This wind generator was designed to be easy to construct, low cost, and suitable for home building. To accomplish this much of the construction was wood and used automobile parts. The paper honeycomb technique of blade construction was refined, using an optimum blade layout program adapted from the paper presented by Ulrich Hutter at the United Nations Conference on New Sources of Energy, and a modern efficient airfoil (the Wortmann foil is commonly used on high performance sail planes).

### Experience

The blade construction technique produced efficient, inexpensive, light weight blades. By putting a wedge under the honeycomb block and tilting the table of the band saw, the cut-out blade is tapered in plan form from tip to root. The flexibility of the paper honeycomb permits twisting after it is expanded on the blade shaft, giving optimum angles of attack along the blade. The extra care and time required to build a blade that is close to optimum aerodynamic shape results in superior performance. With the downwind configuration (blades downwind of the tower, no tail), the blades provide directional orientation. With this design, there was a tendency for the blades to 'walk' around the tower in gusty wind conditions. This has been eliminated by coning the blades in the 12 Footer. Coning is mounting the baldes back at a slight angle, so as the blades rotate they describe a cone instead of a plane. In the effort to keep this design simple it was limited. Due to the lack of an automatic feathering system, it was manually shut down in high winds. The chain and sprocket drive train was noisy and required frequent oiling.

### References


3. **The Aerodynamic Layout of Wing Blades of Wind-Turbines with High Tip-Speed Ratio**, Ulrich Hütter, Proceedings of the United Nations Conference on New Sources of Energy Volume 7, Wind Power, UN Publications, 1964

4. **Windmills: Here's an Advanced Design You can Build**, Hans Meyer, Popular Science, November 1972


## 16 Foot Diameter Solid Foil

### *Description*

*blades*
3 bladed, 16 ft. diameter
Wortmann FX-60-126 profile
paper honeycomb-fiberglass construction

*tower*
30 ft. wood construction, guyed

*other*
automobile differential power transmission

*material cost*
$250

### Design Objectives

This mill was built to experiment with the use of an automobile differential for power transmission. This has been done by a number of different groups including, Brace Research Institute, Volunteers in Technical Assistance (VITA), and New Alchemists (see references). The advantage to this arrangement is that the differential provides a step-up of 3 or 4:1, brings the power to the base of the tower where it can be easily adapted for various purposes, and decreases the overall costs through the use of 'junk' automobile parts.

### Experience

The use of the differential made construction relatively simple. This mill was in operation for only a short time before the blades caught in a rope tied to the tower. In that time, we learned that the drive shaft running to the bottom of the tower had to be sufficiently large in diameter to prevent the drive shaft from 'winding up'. This experiment helped us realize that the problems associated with the design and development of reliable, high speed wind generators are complex and further compounded as the diameter increases.

### References


5. The Design Development and Testing of a Low Cost 10 hp Windmill Prime Mover, R. R. Chilcott, Brace Research Institute Publication no. MT7, 1970

6. Low Cost Windmill for Developing Nations, Hartmut Bossel, Volunteers in Technical Assisstance, 1970

7. **Wind Power**, Earle Barnhart, The Journal of the New Alchemists, 1973


## 15 Foot Diameter Sail Windmill

### Description

This mill was mounted on the same tower and drive train as the 16 foot diameter solid foil. The sails were made of mylar, the spars of aluminum tubing, and rigging of steel cable.

### Design Objectives

This mill was built to test the rigging system eventually used on the 25 foot diameter sail windmill. The characteristics of a sail windmill with low tip speed ratio (2-3 vs. 5-9 for a solid foil) and high solidity ratio (ratio of the blade area to the swept area) make it well suited for driving mechanical equipment directly. It has high starting torque and rotates at a slow speed. This type of mill

is less efficient than a propeller type at higher wind speeds and does not readily lend itself to automatic regulation but can be inexpensive and simple to construct.

### Experience

The mylar was formed into triangular sails with curvature built-in similar to the construction of sail boat sails. The tubing struts and plywood sandwich hub were inexpensive and simple to construct. The cable rigging around the perimeter of the wheel and from a tripod in the center of the hub to the tip of each spar, served to stabilize the spars, reducing their structural requirements. Care must be taken to see that the wind doesn't fill the sails from behind where the rigging isn't available for strength.

### References


8. Construction Plans for Helical Sail, Volunteers in Technical Assisstance, publication no. 11131.1

9. **Windmills of Murcia**, Paul Oliver, Shelter 1973

10. Windmills and Watermills, John Reynolds, Praeger Publishers, 1970


## 25 Foot Diameter Sail Windmill

### *Description*

*blades*
6 dacron sails, mounted on laminated wood spars, 25 ft. diameter

*tower*
42 ft. octahedron module, free standing

*other*
power transmission through automobile differential to base of the tower full power operating range 0-20 mph, with reefed operation in winds exceeding 20 mph system weight excluding tower, 510 lbs.

*materials cost*
$700, excluding sails

### Design Objectives

The sail windmill was designed for use as a prime mover of mechanical equipment with compatible torque and rpm requirements. It was developed as an alternative to the sophisticated construction and materials required for high speed, electric generating 'windmills'. It can be built with minimal shop tools, using locally available materials. The octahedron module tower was developed as a structural system that could be easily fabricated and scaled up or down to meet any tower requirements.

### Experience

The automobile differential drive train was mounted on the tower at a slight angle to equalize disc loading and for better tower clearance. Care must be taken in selecting the correct axle and direction of rotation to insure the proper gear surfaces are being used and the torque exerted on the drive shaft is offset by the axial load on the wheel. The spars can be constructed of wood or tubing, as most of the stresses are taken in the rigging. We used dacron sails, the type used on sailing boats but canvas or other material would be satisfactory. In operation we found the mill self-regulating to a certain extent. If the mill would start to over speed, the sail would 'luff' or spill the excess air. This works for rpm control but is not sufficient to limit the stresses built up in high winds. When strong winds are anticipated the sails are reefed (wrapped around the spars to limit the sail area exposed to the wind). Changing the sail area is also used to match the power available in the wind to the power requirements. The octahedron tower was developed as a free standing tower that would use a minimum amount of materials and would be easy to assemble. In this first generation octahedron tower, the vertical struts of the first module (there are 5) were bent as they passed the first horizontal trangle and continued in to the second module. Smaller tubing fit inside the first tubing and was bent to form the upper portion of module two and the lower portion of module three. The horizontal struts were welded to steel plates at each vertex and the verticals were bolted to these plates.

### References

11. 25 Ft. Diameter Sail Windmill Plans, Windworks, 1974

## 12 Foot Diameter Solid Foil

### *Description*

***blades***
3 bladed, 12 ft. diameter
Wortmann FX-72-MS150B
paper honeycomb-fiberglass construction

***tower***
30 ft. octahedron module

***other***
6:1 step-up, timing belts with jack shaft
12 volt 85 amp alternator
aerodynamically actuated feathering

***materials cost***
$500

### Design Objectives

The 12 Footer was designed with a number of improvements to overcome the limitations of the 10 Footer design. The precision required in construction combined with standard industrial components offers a design solution to the problem of the home-built machine required to run continuously in an open environment.

### Experience

Designed to take advantage of standard industrial components, the basic downwind configuration was kept. Improvements include the feathering-coned hub, timing belt driven train, 30 ft. tower, and slip ring power transmission. A single piece of 2" diameter steel tubing houses the drive shaft which runs to the forward end of the mill and is belted to the alternator via a jack shaft. The entire windmill pivots on a lolly column centered with wooden pillow blocks boiled in linseed oil. This eliminates the need for any bearings for orientation. The alternator used is a 12 volt 85 amp with a 600 rmp cut-in. The timing belt drive train has proved to be quiet, efficient and maintenance free. The tower uses the basic geometery developed for the 25 ft. diameter sail windmill tower, but is constructed of electrical thinwall conduit, an inexpensive source of steel because of its wide spread use. The ends are squashed and drilled for a single bolt at each joint. For testing, the mill was mounted on the back of a flat bed truck, but driving at different speeds did not accurately simulate the suddenness of wind gusts. The experimentation with this mill is on-going. We have load tested the tower to failure and increased the spring constant in the feathering mechanism for smoother feathering operation. Other areas currently under research include a method of limiting the current drain to the field when the mill is not generating and the design of a manual feathering mechanism to shut-down the mill.

### References

12. Construction Plans: 12 Footer, Windworks, 1974
13. Construction Plans, Jim Sencenbaugh, Sencenbaugh Wind Electric, 1974
14. Alternative Sources of Energy, bi-monthly publication on home-built energy systems

## Commercial Wind Generators

There are currently three major manufacturers of domestic-scale, high speed, wind driven generators. They are Dunlite of Australia, Elektro of Switzerland, and Aerowatt of France. They range in size from 50 to 600 watts and in price from $800 to $16,000. There is also a new wave of potential wind generator manufacturers that are trying new concepts and technologies to develop practical, profitable commercial wind systems. Most of these projects are in the prototype stage, and it may be some time before they are commercially available. (references 17, 18).

### References

15. List of Windmill Manufacturers, Gerry Smith, University of Cambridge, 1973

16. Electric Power From the Wind, Henry Clews, Solar Wind Publication, 1973

17. **Wind Power**, Marguerite Villecco, Architecture Plus, May/June 1974

18. **Wind Power: How New Technology is Harnessing an age-old Energy**, E. F. Lindsley, Popular Science, July 1974

Elektro G.m.b.H.
Winterthur
St. Gallerstrasse 27
Switzerland

*distributors:*

Solar Wind Co.
PO Box 7
East Holden, Me. 04429

Environmental Energies
11350 Schaefer St.
Detroit, Mi. 48227

| model | voltage | output | diameter | cost |
|---|---|---|---|---|
| W 50* | 6, 12, 24 | 50 | 1 ft. 5 in. | $ 795 |
| W 250 | 12, 24, 36 | 250 | 2 2 | 1045 |
| WV 05 | 12, 24, 36 | 600 | 8 4 | 1115 |
| WV 15 G | 12, 24, 36, 48, 115 | 1200 | 9 10 | 1430 |
| WV 25 G | 24, 36, 48, 65, 115 | 1800 | 11 6 | 1725 |
| WV 25/3 G | 24, 36, 48, 65, 115 | 2500 | 12 6 | 2020 |
| WV 35 G | 48, 65, 115 | 4000 | 14 5 | 2360 |
| WVG 50 G | 65, 115 | 6000 | 16 5 | 2930 |

* these machines are horizontal axis types

All prices FOB Maine, January 1974 includes feathering, does not include voltage regulator ($660) automatic cut-off, hand brake control, or tower top. They reach rated output at aprox. 25 mph.

Aerowatt S. A.
37 rue Chanzy
75 Paris 11e
France

*distributors:*

Automatic Power Inc.
Div. of Pennwalt Corp.
205 Hutcheson St.
Houston, Tx. 77003
(713) 228-5208

International Aeradio
1165 Leslie St.
Don Mills, Ontario
Canada
(416) 449-3122

| model | voltage | output | diameter | cost |
|---|---|---|---|---|
| 24FP7 | 12, 24 | 24 watts | 3.3 ft. | $ 2,953 |
| 150FP7B | 12, 24 | 120 | 6.7 | 5,429 |
| 300FP7 | 12, 24 | 350 | 10.7 | 7,120 |
| 1100FP7 | 24, 48, 120 | 1125 | 16.7 | 12,809 |
| 4100FP7 | 24, 48, 120 | 4100 | 30.7 | 16,476 |

These prices are FOB Ontario, October 1973, including control panel, excluding taxes.

Aerowatt also offers a FP5 series (vs. FP7) in which the wind generator reaches rated output at 5 m/s (11 mph) instead of 7 m/s (15.5 mph).

Dunlite Electrical Co.
Division of Pye Industries
21 Fromme St.
Adelaide 5000
Australia

*distributors:*

Solar Wind Co.
PO Box 7
East Holden, Me. 04429

Real Gas and Electric
PO Box A
Guerneville, Ca. 95446

Syverson Consulting
2007 Roe Crest Dr.
North Mankato, Minn. 56001

Environmental Energies
11350 Schaefer St.
Detroit, Mi. 48227

| model | voltage | output | diameter | cost |
|---|---|---|---|---|
| "L" | 32 | 1000 watts | 12 ft. | $ 2,025 |
| "M" | 24, 32, 48, 115 | 2000 | 12 ft. | 2,975 |

These prices are FOB Maine, January 1974, includes solid state voltage regulation and control panel with amp and volt meters. This machine is the same as Quirk's, who is Dunlite's Australian agent. These mills reach rated output at 25 mph.

Domenico Sperandio windmills have been used by Italian Radio Television since 1966. They have 7 models, from 100 to 1000 watts.

Domenico Sperandio
Via Cimarosa 13-21
58022 Follonica (GR)
Italy

Dyna Technology makes 6 and 12 volt Winchargers.

Dyna Technology
P.O. Box 3263
Souix City, Iowa 51102

Lubing is a German firm that manufactures electric and water pumping windmills. The electrical wind generator is a 400 watt, 24 vdc system.

Lubing Maschinenfabrik
Ludwig Bening
2847 Barnstorf (Bez. Bremen)
Postfach 171
Western Germany

### Tip Speed Ratio

$$u \text{ (tip speed ratio)} = \frac{2\pi R N}{V}$$

where:
R = radius (ft.)
N = revolutions per minute
V = wind velocity (ft./sec.)

The tip speed ratio is inherent in the design of a given type of windmill depending on the blade efficiency, ratio of blade area to swept area, and operating conditions. For windmills designed to operate in light winds with a high solidity ratio, the tip speed ratio tends to be low; for mills that operate efficiently in high winds and rotate at high speed the tip speed ratio can be as high as 13. (see chart)

### Power Equation

$$P = \tfrac{1}{2} A q V^3 Cp$$

where:
P = power
A = area
q = density of air
V = wind velocity
Cp = power coefficient

Assuming sea level air density, this can be rewritten as:

$$P = K A V^3 Cp$$

Where the value of K for different units of power, area and velocity are:

| Power | Area | Velocity | K (constant) |
|---|---|---|---|
| kilowatts | sq. feet | miles per hour | $5.3 \times 10^{-6}$ |
| horsepower | sq. feet | miles per hour | $7.1 \times 10^{-6}$ |
| watts | sq. feet | feet per second | $1.68 \times 10^{-3}$ |
| kilowatts | sq. meters | meters per second | $6.4 \times 10^{-4}$ |

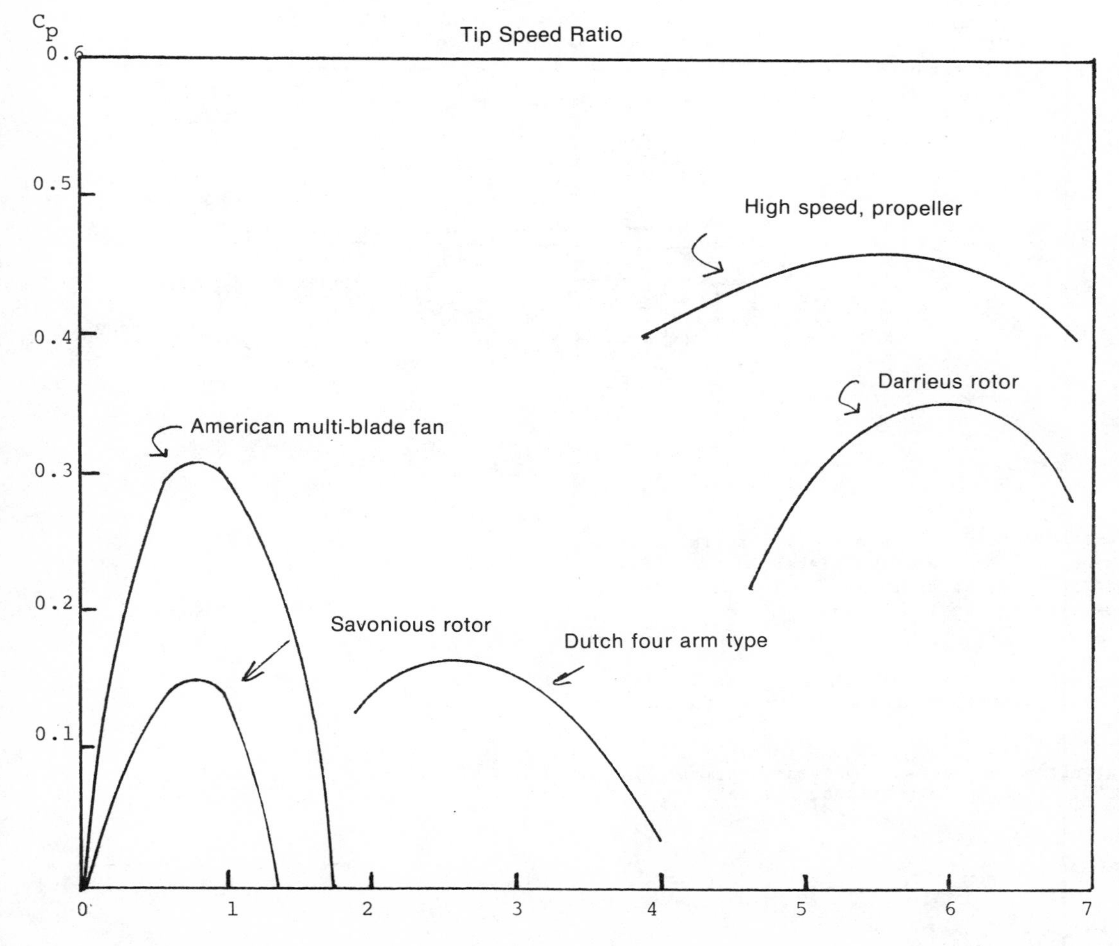


The power coefficient (Cp) is a non-dimensional number that reflects the amount of energy a given windmill design can extract from the wind. The theoretical maximum is .593 that is, a windmill that is 100% efficient would have a power coefficient of .593 but in reality that is never attained. The power coefficient, or efficiency of any design is dependent on many different factors, but some approximate maximum values for various design are given below.

**Savonious Rotor: .15**
This is a vertical axis machine, often made with 55 gal. drums.

**Dutch four Arm: .17**
This horizontal axis mill has been used historically for grinding grain and pumping water.

**Sail Windmill: .3**
This is the design we have worked with, it was originally developed in the Eastern Mediterrian.

**American Multi-blade Fan: .3**
This design was used extensively in the West and Middle West.

**Darrius Rotor: .35**
This is a vertical axis machine, often described as the "egg beater" design.

**High speed propeller: .45**
Two or three bladed, commonly used for the generation of electricity.

### References

19. Is There a Place for the Windmill in the Less Developed countries?, M. Merriam, Technology and Development Institute, Working Paper no. 20, 1972

20. **Windmills**, E. N. Fales, Standard Handbook for Mechanical Engineers, McGraw-Hill, 7th ed. 1958

21. Applied Aerodynamics of Wind Power Machines, R. E. Wilson and P. B. S. Lissaman, NSF grant study, 1974

### Wind Data

A climatic atlas of the U.S. ($4.25), or a monthly report for each state ($2.40/yr. or 20¢ ea.), both with wind averages, strongest wind, and wind direction are available from:

Environmental Data Service
National Climatic Center
Federal Building
Asheville, N.C. 28801

Wind speed measuring equipment, including Dwyer and Taylor anemometers, are available from:

Aircraft Components
North Shore Drive
Benton Harbor, Mi. 49022

### Brace Research

Brace Research Institute works in the field of intermediate technology for developing countries. They have 360 publications, largely on solar energy, 30 on wind. A description of their 32 foot diameter windmill is in "The Design Development and Testing of a Low Cost 10 hp Windmill Prime Mover" by R. E. Chilcott. (publication no. MT 7)

Brace Research Institute
MacDonald College —
McGill University
Ste. Anne de Bellevue 800
Quebec, Canada

### Information Sources

The Electrical Research Association, with 25 years experience in the field of wind energy, has 35 publications including many translations from Russian literature.

Electrical Research Association
Cleeve Road
Leatherhead Surrey
England

NSF-NASA sponsored a conference on wind driven generators in June 1973. Primary focus was on installations large enough to supply 10% of the nation's energy. Proceedings, with a description of NASA's wind energy program are available from:

NASA
Lewis Research Center
Mail Stop 500-201
Cleveland, Ohio 44135

"Is There a Place for the Windmill in the Less Developed Countries" is a paper covering the basic aspects of windmills, available from:

The East - West Center
Honolulu, Hawaii 96822

# Carving Your Own Propellor

*From Alternative Sources of Energy #8*

By Winnie RedRocker

This is really the heart of the thing—and if nothing else, the main point of this article is that a wooden propellor of this type is *easy* to build with almost no woodworking skill and very little money. The plan I used is from the LeJay Manual and I've included step-by-step instructions which I hope anyone can follow to make a prop, using a good piece of wood, a drawknife, a rasp, and some sandpaper. If you've never seen a high-speed prop going in the wind, by all means try building this one. Mount it on an old generator or something and set it up in the wind. Don't hold it in your hands!! I almost lost my head that way. It'll blow your mind when you get a feeling of the power that this thing produces!

*Winnie RedRocker*
*PO Box #3*
*Farasita, Colo. 81073*

**STEP 1:**

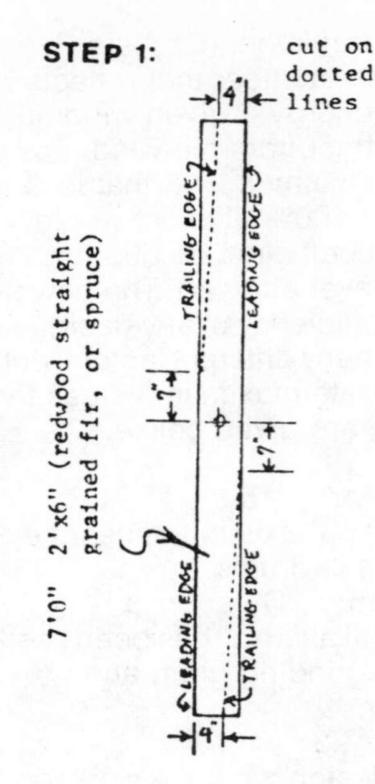


**STEP 2:**

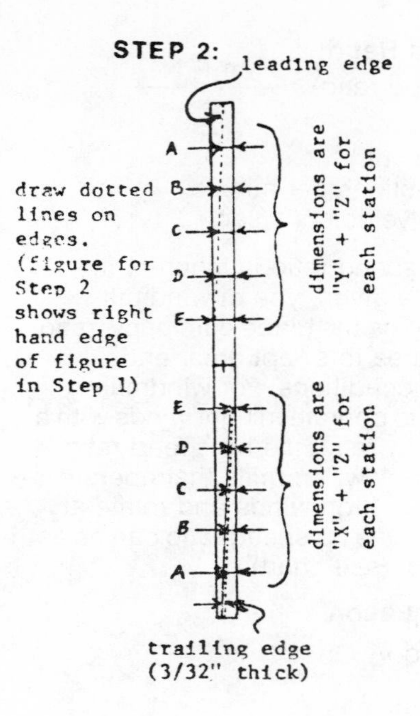


**STEP 4:**

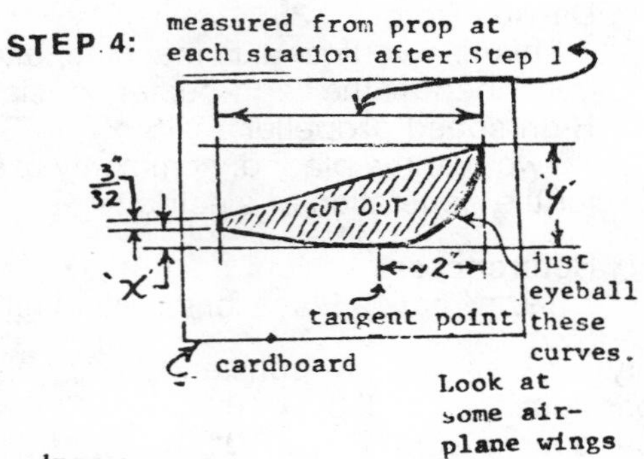


draw cross-sections for each station full size on cardboard and cut out with razor to make templates.

**STEP 3:**

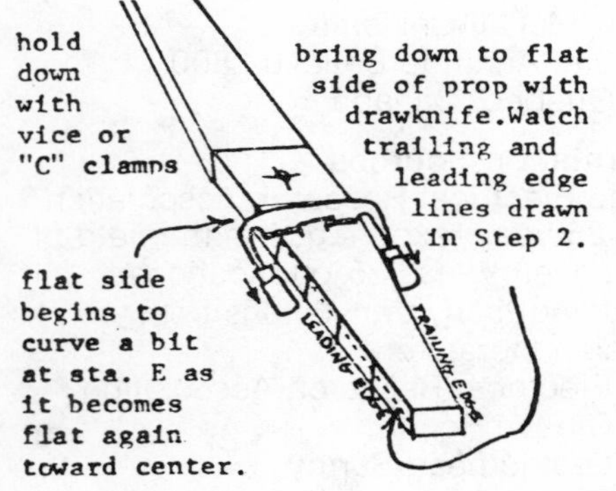


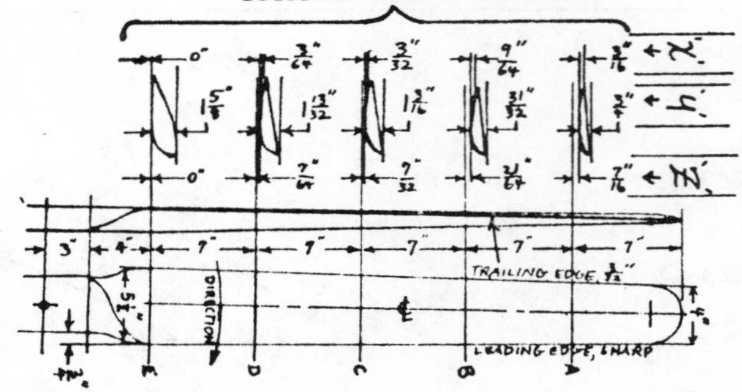


**STEP 5:** start at tips

bringing wood down on airfoil side of prop with drawknife(or rasp when it starts getting close) until template for station A will fit on prop in the right place (7" from tip) -- then go to template B until it fits (7" from A) -- and so on toward the center. The flat side made in Step 3 is your reference for positioning the templates.

**More Steps:**

sand well, varnish (several coats), and balance both horizontally and vertically. Balancing can easily be accomplished by adding small weights to the edge (vert. balance) and front (horiz. balance).

NOTE: This prop can be made 10 ft long by extending the distance between stations from 7" to 10", making the 3" x 4" sections both 5". All other dimensions, the same.

# Balancing The Blades

By Ted Ledger

*from Alternative Sources of Energy #10*

I have some good news—a set of blades in good static balance are also in good dynamic balance. The reason is that blades are so thin in section that there is no room for axially displaced concentrated weights. In a car wheel (see Figure 1) and tire, there's an axial width of 6″ or more, but a windmill blade is less than ½″ front to back, and made of very uniform carefully chosen wood, or other uniform material.

Good balance means mounting the blade on a mandrel and letting it roll on a couple of carefully levelled edges—like angle iron set this way. (Figure 2.) Make sure the sharp edges are really sharp. If necessary, file carefully. Only about ½″ is necessary.

Load the light blade by hanging pieces of wire or solder out near the tip. Let this be done on a windless day or inside a building. Then drill a hole in the middle of the tip section, to meet in the metal. (Figure 3.) If more than one hole is needed, make holes as symmetrical as possible.

After balancing, the next job is to "track" the blades. Mark one blade with a coat of bright paint, and coat it all over. Then mount it and allow it to run free—or with a generator load, as wished. Watch the blades as they spin. If the pitch of each is o.k., the blades make a single line, running, viewed from the edge. If one is behind the other (in edge view), that blade has too little pitch compared to the other. Adjust by putting a shim under the blade-hub mount to increase the pitch of the "too flat" side. (Figure 4.)

Try it again in the wind. The colored blade helps to identify one from another in deciding which is low pitch, which is high. This assumes a two-blade windmill. Once the blade tracks right, your job is done.

If you have a multi-blade mill make a dummy mount to hold two blades only. Balance all the blades against the heaviest, and for a check, try a few of the balanced blades against each other in the dummy two-blade mount. Hope this is useful!

*Ted Ledger*
*591 Windermere Ave.*
*Toronto, Canada*
*M6S 3L9*

**Figure 1**

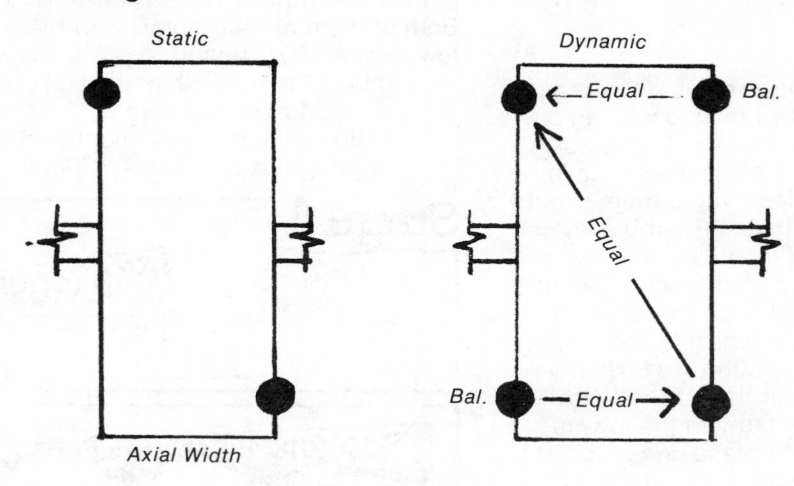


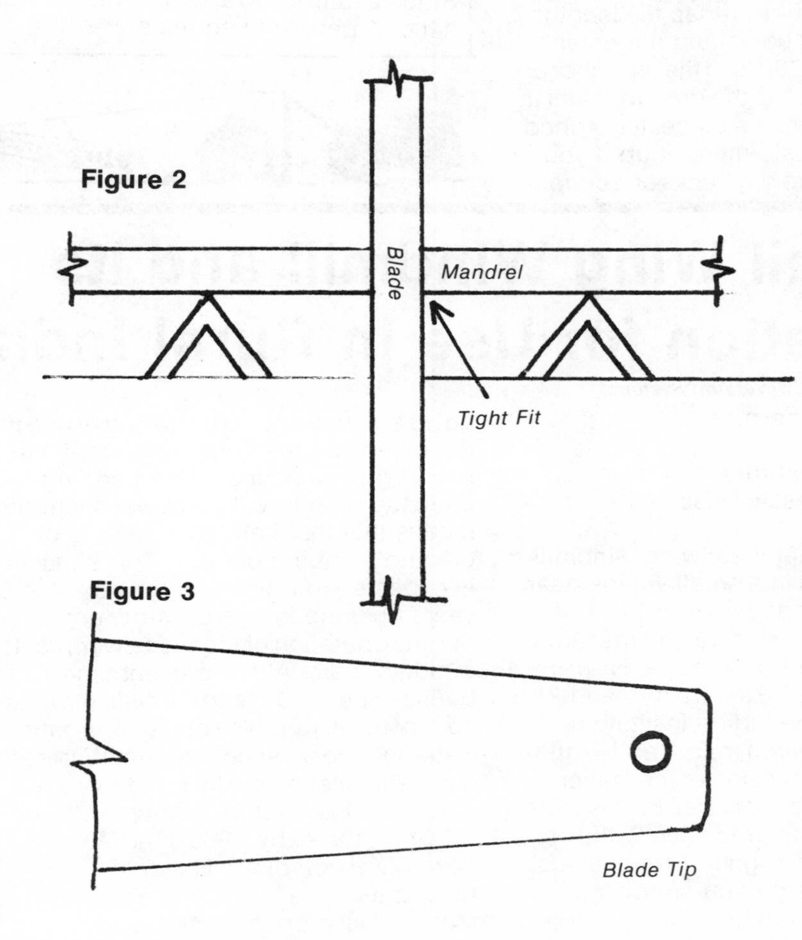


**Figure 2**

**Figure 3**

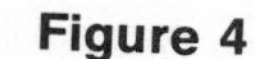

**Figure 4**

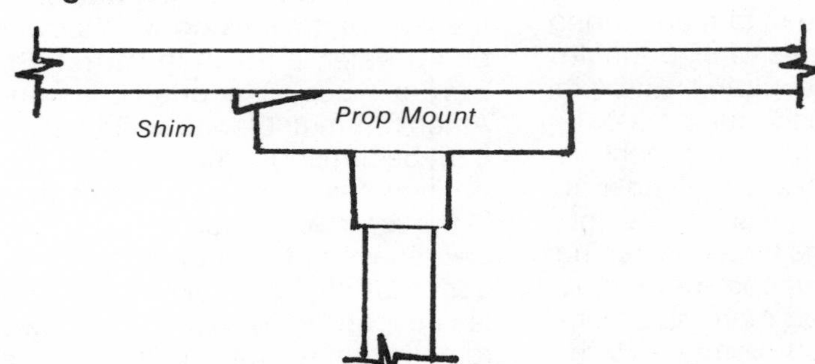

# Towers

10/15/74

Dear John,

Let me fill you in on moving a farm type windmill. The one I moved was an old one maybe 60 years old. And it was riveted together. Unless the mill is to be moved a few miles riveted mills should be avoided. I had to disassemble my mill and transport it 200 miles, so I had to remove rivets in certain places and use bolts when I put it back together. Removing rivets is a hassle and buying bolts is expensive. Mine is a 50 ft. F & W from Kendalville, Indiana. From what I've heard from some farmers the towers usually are built in place one piece at a time. Well, this isn't so bad if the mill is bolted together but if you have a situation like mine where I had to shorten the mill about 2 ft. so I could overlap the lengths of angle, drilling holes and alignment would be a hassle 30 ft. in the air. I chose to assemble mine on the ground, paint it and then raise it in one piece. It weighed 1000 lbs. No hassle lifting it up if you have access to two big trucks or a couple of tractors. See sketch.

I have located 2 other towers which I plan to take down around Thanksgiving. Both of them are Air-motors and the towers are 100% bolted together, so it should be a lot easier than the first one.

R. Todd Swinderman
7804 Johnsburg Wilmot Road
Spring Grove, Illinois 60081

**Stage 2** Lift mill to approximately 20° by hand. 4 people lifted mine.

**Stage 3** Using a gin pole or in my case an end loader, pull from a height above the mill (attach rope about in the center) and raise until 2nd vehicle can lower it. My mill didn't reach the turning point until about 75° then the 2nd truck lowered it slowly. This works way better than trying to push it up from the underside. I used ½" rope and it held.

# The Sail Wing Windmill and its Adaptation for Use in Rural India

*From Wind Energy Conversion Systems*

By Marcus M. Sherman

New Alchemy Institute
Woods Hole, Massachusetts

A 25-foot diameter sailwing windmill was built in 1973 in a small village near Madurai, Tamilnadu State, India. This windmill is the result of design research conducted in the U.S.A. at the New Alchemy Institute—East (refs. 1 and 2) and in India at the Indian Institute of Agricultural Research (refs. 3 and 4) and the Wind Power Division of the National Aeronautical Laboratory (ref. 5). It is to be used mainly in light winds during the dry winter months for irrigating small fields, watering dairy cattle and supplying water for domestic use.

In many parts of India there are adequate supplies of ground water which are unavailable to farmers during the dry season because of inadequate power resources for pumping. Three to eight horsepower diesel pumps are frequently used, but they are expensive to operate because of the high cost of imported oil and often must be taken out of service for costly and time-consuming repairs. Efficient 5-horsepower electric pumps are being used more and more are rural electrification proceeds, but only well-to-do farmers can afford to buy and maintain them. Recently in South India there has been a 75 percent power cut to the rural areas due to heavy use in the cities and to overexpansion of the power grid without a corresponding increase in supply. This power shortage means that there are only 4 hours of electric pumping per day. This situation is expected to worsen for the next 4 to 5 years until the Indian Government begins operation of atomic power plants in South India. At the present time bullock operated pumps remain the most common and reliable source of irrigation water for subsistence farming. Water for domestic use is usually hand-lifted with a rope and bucket from open wells.

During the early 1960's the Wind Power Division of the National Aeronautical Laboratory in Bangalore, Mysore, developed, tested, and produced two hundred 12-bladed fan-type windmills which demonstrate the feasibility of using wind power to pump water to South India (ref. 6). Several types of imported European and American multibladed windmills have also been used to harness India's abundant wind energy resources. However, due to lack of public awareness of the subject and the unavailability of an even simpler and less expensive device, wind power remains only occasionally exploited.

Cloth sails with a wooden framework have been used for hundreds of centuries for transforming the useful energy of the wind into labor saving mechanical work, especially grinding grain and pumping water. The use of windmills spread from Iran in the seventh century A.D. to coastal China where the application of the art of sailmaking significantly improved the sophistication of windmill construction (ref. 7). Heavy rigid wood mindmill blades surfaced with cloth were increasingly used throughout northwestern Europe so that by the seventeenth century the Netherlands became the world's richest and most industrialized nation, largely as a result of extensive exploitation of windpower with ships and windmills. Cloth was a natural choice for windmill sails because of its acceptance and wide use in sailing ships. It is lightweight, easy to handle, readily and cheaply available, and most importantly it forms a strong uniform surface for catching the wind when firmly supported at three or more points.

In the Mediterranean region flour-grinding and oil-pressing mills were rigged with six to twelve triangular cloth sails set on simple radial spars. A three-dimensional array of guy ropes radiating from a central spar projecting out along the axis of the main shaft suspended the sails in position, rather than a heavy grid of wood as was used in the traditional Dutch-type windmills. This sailboat jib type of rigging was a significant improvement in windmill design which encouraged the spread of windmills throughout the deforested Mediterranean countries. The wind capturing area of these windmills was controlled by wrapping each cloth sail around its spar. Though requiring daily rigging adjustments and occasional replacement of tattered sails, the efficiency and simplicity of these windmills resulted in their widespread use in Rhodes, the Black Sea coast, the Aegean Islands, and Greece. In Portugal their use was accompanied by the sound of whistles attached to the rigging, an audible indicator of the wind at work. In the West Indies large sailing windmills were commonly used for crushing sugar cane (ref. 8). Many handcrafted windmills with eight triangular jib sails are presently pumping irrigation water in the Plain of Lassithi, Crete (ref. 9). In Japan four-bladed jib sail windmills are used to operate reciprocating pumps which supply water to vegetable gardens. A high-speed aerodynamic, two-bladed sail wing is being developed (refs. 10 and 11). Further construction simplifications may make it applicable to use in lesser developed countries.

A windmill with four self-adjusting cloth sails was developed for rural markets in less industrialized regions (ref. 12). Its relatively complex design is limited because of the difficulty in connecting it to a deep well pump. Unfortunately, it cannot be manufactured by hand using local materials. Those people who are in a situation to most benefit from a windmill are also those least able to pay for it. If the critical moving parts were separately available, a small farmer could purchase the remaining materials needed and assemble the windmill in his own village using local skills and labor. This way a major portion of the money spent would remain in the village.

The 8-meter-diameter prototype sail wing windmill recently erected on a small peanut and sesame farm in a dry hilly region in South India lifts 300 pounds to a height of 20 feet in 1 minute in a 10 mph wind. This is accomplished by a rope passing over a 6-inch pulley on the main drive shaft. This lift is used to lift soil and rock from the well being hand dug below the windmill. The windmill will be set up to operate a modified paternoster or chain pump like those used to drain mines in England many years ago. Recently chain pumps have been rapidly replacing the traditional square-pallet pump and the noria water lifting wheel throughout China. A chain pump, easily and cheaply built, is more efficient than most types of pumps. Most importantly, it operates well with a low-speed, variable power source.

This sail wing windmill is made of a 1-meter-diameter bullock cart wheel to which three bamboo poles are lashed in a triangular pattern with overlapping ends. Each bamboo pole forms the leading edge of a wing, and a nylon cord stretched from the outer tip of the pole to the rim of the wheel forms the trailing edge. A stable and lightweight airfoil results from stretching a long narrow triangular cloth sail over that bamboo-nylon frame. This wing configuration, a hybrid of low-speed eight-bladed Cretan sail wings and high-speed two-bladed aerodynamic sail wings, produces high starting torque at low wind speeds. The bullock cart wheel is attached at the hub to the end of an automobile axle shaft which rotates in two sets of ball bearings. The shaft and bearing assembly is mounted horizontally on top of a turntable. The turntable consists of two circular steel plates separated with a raceway of ballbearings and held together with a ring of eight bolts which encircle the bottom plate. A 1-foot diameter hole through the center of the turntable will allow the chain and gaskets of the chain pump to go up and around the "squirrel cage," which is mounted at the center of the automobile axle. If a reciprocating, deep-well piston pump were desired, the reciprocating rod, rather than a chain, would be mounted on top of the turntable. Since the blades have a slight built-in coning effect and the axle or crankshaft is mounted slightly off center from the centerline of the turntable, the blades act as their own tail, trailing in the wind. Because the blades are downwind from the tower, there is no danger of the bamboo poles bending in a monsoon wind and hitting the tower. The tower is made of five 25-foot long teak poles set in concrete at the base and bolted at the top to five angle irons welded at a slight flaring angle to the bottom of the turntable. The tower tapers in towards the turntable at the top from a 7-foot diameter at the base. It has cross bracing and a ladder.

It is hoped that other persons will continue to refine and adapt this windmill to their own needs and materials. Please send all inquiries, operating experience, and suggestions for improvement to: Marcus M. Sherman, New Alchemy Institute—East, Box 432, Woods Hole, Mass. 02543.

### References


1. Eccli, Eugene: Alternate Sources of Energy Magazine, Mar. 1973, page 17. Box 73, Kingston, New York 12401.
2. Sherman, Marcus M.: A sail Wing Windmill in India. J. New Alchemists, New Alchemy Institute, Woods Hole, Mass., 1973.
3. Neermal, T. H.: Indian Institute of Agricultural Research, Puse Institute, Machinery Development, New Delhi: private communication.
4. Michael, A. M.: Indian Institute of Agricultural Research, Pusa Institute, Water Technology Center, New Delhi: private communication.
5. Gupta, Sen: National Aeronautical Laboratory, Wind Power Division, Bangalore, India: private communication.
6. Harness the Wind with the WP-2 Windmill. Published by Wind Power Division, National Aeronautical Laboratory, Bangalore, India.
7. Needham, Joseph: Science and Civilization in China, vol. IV, Mechanical Engineering. Cambridge University Press, Cambridge, England, 1965.
8. Chilcott, R. E.: Implications of the Utilization of Wind Power for the Development of Small Caribbean Communities. Brace Research Institute, McGill University, Montreal, Pub. No. R25. 1971.
9. Reynolds, John: Windmills and Watermills. Praeger Publishers, New York. 1970.
10. Kidd, Stephen and Garr, Douglas: Electric Power from the Wind. Popular Science Magazine, Nov. 1972.
11. Fink, Donald E.: New Air Foil Design Method Developed. Aviation Week and Space Technology, Nov. 1972.
12. Proceedings of the United Nations Conference on New Sources of Energy, Rome, Aug. 1971, vol. 7, Wind Power. United Nations, 1964.


### Further Reading


Merriam, Marshal F.: Is There a Place for the Windmill in the Less Developed Countries? Working Paper Series No. 20, Technology and Development Institute, East-West Center, Honolulu, 1972. Also: Windmills for Less Developed Countries. J. Intern. Division of the American Society of Engineering Education, TECHNOS, Apr.-Jun. 1972.

Merriam, Marshal F.: Decentralized Power Sources for Developing Countries. International Development Review, vol. 14 no. 4, 1972. Also: Working Paper Series No. 19, Technology and Development Institute, East-West Center, Honolulu, 1972.

# Experience with Jacobs Wind-Driven Electric Generating Plant, 1931-1957

*From Wind Energy Conversion Systems*

By Marcellus L. Jacobs
Jacobs Wind Electric Company, Inc.
Fort Meyers, Florida

This report outlines the engineering, construction, performance, electric output, and different uses of the Jacobs wind electric 2500- to 3000-watt plant, thousands of which were installed in many parts of the world between 1931 and 1957.

Early engineering started on this wind-operated electric generating plant in 1925. After several years of testing different types of windmills, the three-blade aeroplane type of propeller was found to be far superior in power output. By means of a flyball-governor-operated, variable pitch speed control, the maximum speed of the propeller was accurately and easily controlled, to prevent excessive speeds in high winds and storms. The three-blade propeller was found to be necessary (as compared to the two-blade type) to prevent excessive vibration whenever the shift of the wind direction required the plant to change its facing direction on the tower.

The periods of vibration which occurred on the two-blade propeller, every time the tail vane shifted, to follow the changes in wind direction, were found to be caused by the fact that the two-blade propeller, when in a vertical position, offers no centrifugal force resistance to the horizontal movement of the tail vane in following changes in wind direction. However, when the two-blade propeller is in the horizontal position, its maximum centrifugal force is applied to resist horizontal movement of the tail vane; thus the tail vane is forced to follow wind direction changes by a series of jerks, causing considerable serious vibration to the plant.

The three-blade propeller was developed by us in 1927 to correct this condition. When in operating, the three-blade propeller creates a steady centrifugal force resistance, against which the tail vane reacts with a constant pressure and produces a smooth shifting horizontal movement of the plant facing direction. The centrifugal force generated by the very light aeroplane spruce-wood blades, when operating at 225 rpm is 550 pounds each, making a force of over 1600 pounds of gyroscopic resistance force to the horizontal vane movement for the three blades. But this resistance is in the form of an even pressure or resistance to horizontal movement, whereas the 1100 pounds of gyroscopic resistance force of the two-blade propeller to the vane movement is applied and then eliminated twice during each revolution.

A propeller diameter of 15 feet was found to produce ample power for electric generator operation to develop 400 to 500 kilowatt-hours per month, based on the available winds in most areas of the states in the western half of the United States. This required 10 to 20 mph winds for 2 or 3 days per week. A specially designed six-pole battery charging type shunt generator was developed to operate at a speed range from 125 to 225 rpm for direct connection to the governor hub of the propeller. It was designed so that its load factor would exactly parallel the power output curve of the wind-driven propeller when operating in the 7 to 20 mph range that was felt to produce the most hours of wind per month. Wind plants that require higher than 20 mph winds to deliver their rated output will find too many areas where there are too many days with winds below that speed each month, and thus their effective average monthly output in many areas is below expectations. The generator weighs 440 pounds with a 9-inch-diameter armature with a 9-inch core length. The 60 pounds of wire on the field poles gave maximum efficiency with a drain of less than 100 watts for field coil operation. The generator output is 2500 watts at 32 volts, and, for the 110-volt generator, it is rated at 3000 watts.

Our experience with plants installed in many parts of Alaska, Canada, Finland, northwestern United States, and a number of special installations such as the plant we have installed for the joint operated United States and United Kingdom weather station at Eureka, in the Arctic Circle, and with the Byrd Expedition at Little America has shown that aluminum painted (copper edged) spruce-wood propellers have considerably less trouble with frost and ice formation than when they are varnished or when other type coatings are used.

Generators located on high-steel towers are subject to considerable static discharge from the armature through the ball or roller bearings, and excessive charges from nearby lightning will often arc through a bearing and weld spots on the balls and race, causing it to break up soon. We found the revolving propellers collected discharges into the direct connected armature and the lightning pick-up effect of the propellers was frequent and of considerable intensity. To correct this, we installed dual sets of heavy grounding brushes on the armature shaft which completely eliminated any trouble from this cause. With the additional use of a large capacity oil-filled condenser connected across the generator brushes and frame, we practically eliminated any damage to the generators from lightning, so much so that, with high grade ample insulation used throughout the generator and the grounding brushes and condensers, we gave an unconditional 5-year guarantee with every generator against burn-out from any cause and have built many thousands during the past 20 years using this construction without any replacements ever being required because of lightning damage or burn-out from any cause.

The price received at the factory for our 2500-watt, 32-volt plant was $490, less the cost of a suitable tower and batteries, which could often be secured in the country or area to which the plant was shipped. We supplied a 21000-watt-hour glass cell lead-acid type of storage battery with a 10-year guarantee, for which we received $365. A fifty-foot self-supporting steel tower was supplied for $175, making a total cost for the plant of $1025. This is about $400 per kilowatt as the manufacturing cost of the plant. Shipping and installation costs are additional. Installation cost requires only the labor of two men for two days and a small amount of cement to put into the anchor holes when the tower is built. No special equipment or training is necessary. We have shipped hundreds of plants to most countries with not a single request for additional information to enable them to erect the plant. Regular installation and operating instructions are prepared and sent with each plant.

Operating and maintenance costs of this plant are largely limited to the replacement of the storage battery which, on a 10-year basis, is about $36 per year; from records kept of more than 1000 plants over a 10-year period, the maintenance cost of repairs was less than $5 per year. Some of the owners of our plants bought the Edison type battery and after 20 years are still using the same battery. New batteries of this type are quite expensive, but these owners bought second-hand batteries which still gave them 20 years of service.

Special generators designed for the cathodic protection of underground steel pipelines were developed by us in 1936. These generators were wound for an external circuit resistance of 1/10 ohm or higher. The generators produced 10 volts at 100 amperes and were straight shunt wound. When connected to the pipelines in any normal wind, they maintained a pipe-to-soil potential of 3/10 of a volt pipe negative. Due to the action of the current, the pipe maintained a fair degree of protection through calm wind periods. Hundreds of our plants are protecting many miles of pipelines in North and South America and in Arabia. Some of these plants have been in service since 1937.

### Discussion

Q: Can you tell me the present state of this design? You say you are no longer manufacturing wind generators, but are the designs available?

A: Well, I closed the plant and sold the machinery. I still have the company, but the engineering I do is a different type of engineering now.

Q: Are these designs available if another company is interested in producing it?

A: Frankly, it's been 18 to 20 years since I last produced wind generators, and I haven't made much effort to keep them. I'm busy with environmental work, developing a system for cleaning up coastal canals and waters (I have patented and developed a system for that), so I have dropped out of the wind electric business. Now, there are a lot of old plants still running here and there around the country, but no new ones. I no longer have the plans, blueprints, or information on them. I didn't keep them.

Q: Would you have any guess as to what these units would cost today in per kilowatt?

A: They would be about twice what they were when we quit building them.

Q: That's a complete system?

A: That's the plant, tower, and suitable storage battery.

Q: We had earlier a very interesting discussion on the question of electric plants. It would be of interest if you could comment on the operation of such gear.

A: Early in the thirties, about 1931 or 1932, I made a series of tests, and we put a special grounding brush on the generators. We have found that the airplane spruce propellers with the copper leading ends and the static pickup, out in wind and sand and from certain atmospheric conditions, created a static buildup in the armature, which would jump across to the main frame through the ball bearings and would wreck and damage the bearings.

And then I discovered in 1932, that by putting a set of heavy grounding brushes on the big armature shaft, which is 2 inches in diameter, that eliminated that completely. After that no bearings ever went bad and there was no more static buildup.

# Alternators and Their Uses

No matter what form of energy we start with, we usually end up by trying to convert it into electricity. Electricity is one of the easiest and most convenient ways to handle and use energy. When the problem of generating electricity with windmills, waterwheels, etc. is under discussion, the means most often suggested is the automobile alternator.

Alternators are plentiful and inexpensively obtained from junkyards. Of course you must be careful to pick one that is in good condition. Be sure to find out the make and model of the car that the alternator was used in. With this information, you can look your find up in an auto repair manual. There are variations in specifications and hookups, and it is wise to be sure that you are using the alternator correctly. The manuals will also suggest tests that you can use to spot defects.

Alternators produce 12 volts direct current and are commonly available in current ratings ranging from thirty to sixty-five amperes. The power available from an unmodified alternator is rather limited when compared to the sort of power we are used to using. A 65 amp. generator at full load puts out 12 × 65 watts or 780 watts—slightly more than three quarters of a kilowatt.

The basic principle behind all alternators and generators is that an electric current flows in a loop of wire when it moves through a magnetic field. One way to generate electricity is to spin a coil of wire in the field of a couple of permanent magnets; another way to do it would be to let the coil of wire stand still and spin the magnets. The alternator uses the second idea. The magnetic field is developed by an electric current flowing in a coil of wire wound on the spinning shaft. A stronger field is possible this way than you can get with magnets. Logically enough, the winding that produces the field is called the field winding. Current to it is supplied by brushes and slip rings.

The output of the alternator is taken from windings wound through slots on the inside of the frame. These windings are stationary so they are called stator windings (just as the field winding is sometimes called the rotor winding). There are three stator windings. They produce alternating, not direct current. The windings are arranged so that when the current in one is increasing, the current in the second has reached its peak, while the current in the third is decreasing. This is three phase alternating current, and while it is convenient to make motors and generators this way, batteries need direct current. The AC is changed to DC, or rectified, by a set of six diodes. A diode will allow current to pass through it in one direction, but will block current trying to go in the other direction. If you take the flow of current to be from positive to negative, the arrowhead of the diode symbol indicates the direction in which current must flow.

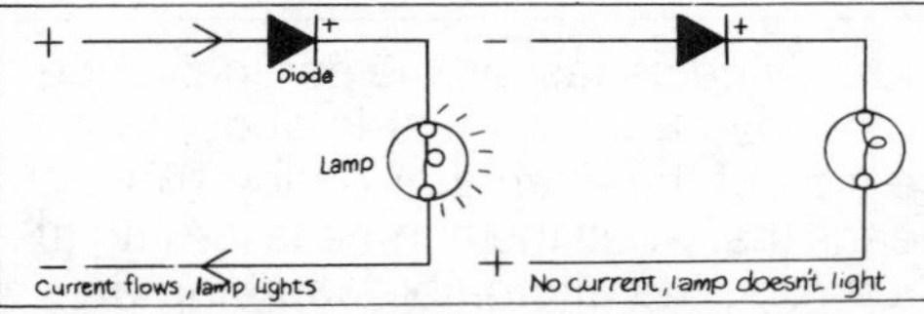


Pick any two stator windings and imagine current flowing out of them, while current flows into the third. Follow the paths through the windings, and you will see that the diodes route the current so that it must flow out of the positive terminal and into the negative terminal. The outputs from the three windings overlap to produce a fairly constant DC voltage. The diodes get slightly warm in operation. To prevent the heat from building up and damaging them, they are pressed into the metal case of the alternator so that the bulk of the metal will carry away the heat.

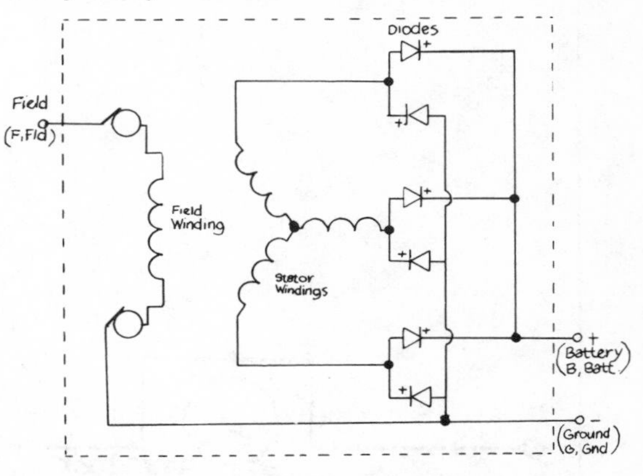


Here's an idea that may be of interest. Since an alternator is really an AC generator, that means that we could get 12 volts of ordinary single phase AC out of it by disconnecting the diodes and using only two of the stator windings.

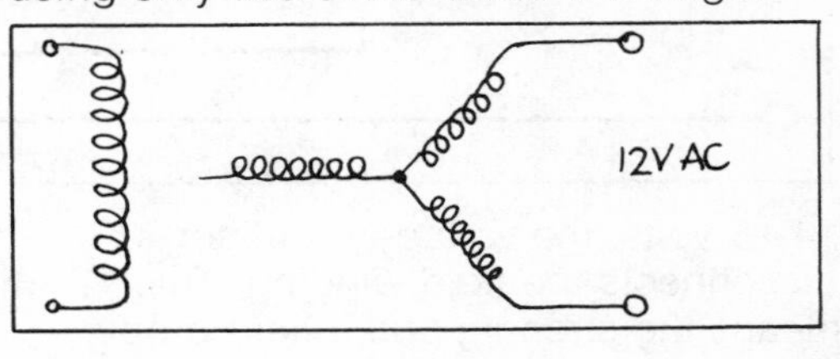


Ordinary house current alternates, or changes direction, 60 times per second. The alternator produces AC with a frequency of several thousand cycles per second, depending on the speed of the alternator. The faster it spins, the higher the frequency of the output.

The output of an alternator depends on two things: its speed and the strength of the magnetic field produced by the field winding. The faster it spins, the more current it can put out, and the stronger the field, the greater the output voltage. A battery can be damaged if it is charged at too high a voltage, so the voltage of the alternator must be kept within bounds. This is the job of the voltage regulator. When the alternator output rises over 14-15 volts, the voltage regulator disconnects the field winding. This means the strength of the field drops, and with it, the alternator voltage. When the voltage has fallen back to safe limits, the regulator allows current to flow through the field coil once more. In this way, the regulator keeps the voltage reasonably constant. Some cars have transistorized voltage regulators. These do a much better job because they can adjust the current through the field coil to just the right value, instead of being limited to full on or full off as the older regulators are. Be sure that you have the proper regulator to go with your alternator.

Most alternators do not begin to produce any significant output until they are turning at 500-1000 rpm. They produce their full rated output at about 2000 rpm. In a car, the alternator turns about twice as fast as the engine, since the pulley on the engine is about twice the size of the alternator pulley. This means that when the engine is turning at 4000 rpm, the alternator is turning at 8000 rpm. The ability of the alternator to stand high rpm without flying apart is one of the reasons that it has become standard on practically all modern cars.

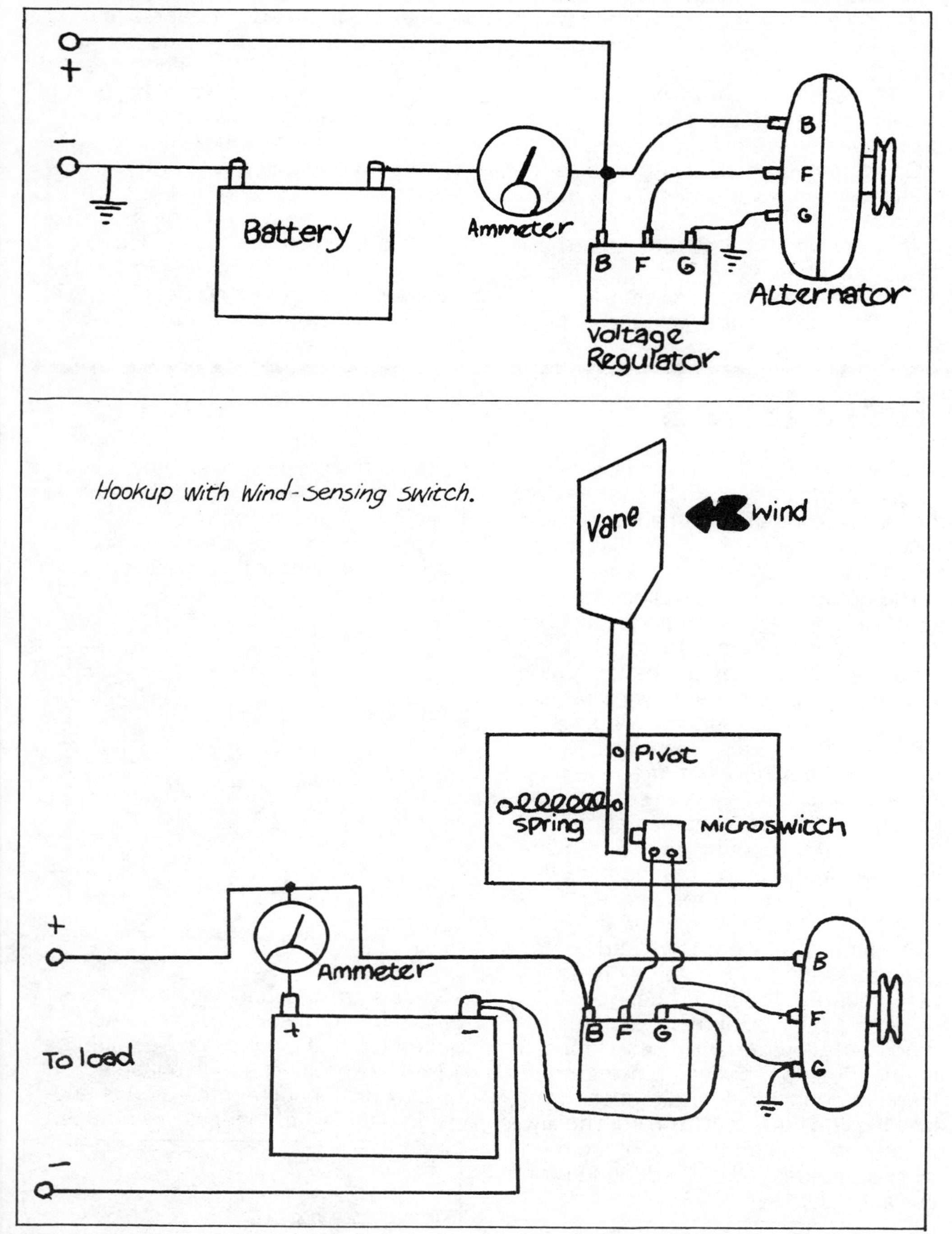


One problem with using an alternator in a wind powered system is that the field winding continues to draw current when the alternator is not turning. This can completely discharge the battery during a long calm. What is needed is a way to switch in the field coil only when the alternator is turning fast enough to generate current. One way might be to mount a centrifugal switch on the alternator shaft. An easier way is to use a switch connected to a spring loaded wind vane. The spring tension is adjusted so that the switch closes when the wind is blowing hard enough to turn the alternator fast enough to charge the batteries. The switch is placed in the wire between the field connection of the regulator and the field connection of the alternator.

An alternator that is mounted high on a tower as part of a wind generator becomes a target for lightning. Even a nearby strike can produce enough voltage in the wiring to destroy the diodes in the alternator. To reduce the lightning hazard, one side of the circuit should be connected with a length of heavy wire to a cold water pipe or a metal rod driven into the ground. This is the side (usually negative) that would normally be connected to the frame of the car. The connection can be identified by the marking "G" or "ground" stamped into the terminals of the alternator and voltage regulator. When grounded, the system has the same sort of protection that would be given by a lightning rod.

Once you have an alternator set up and working, the question becomes how to make use of its output. This is a lot easier than it used to be, now that so many 12 volt DC appliances are being made for trailers and recreational vehicles. Use heavy wire (#12 or heavier) in making your hookups, and don't make the wires any longer than they have to be. At this low voltage, the electrical resistance of the wire can be significant, and it is possible to lose a lot of power that just goes into heating up the wiring.

Most solid state electronic gear does not run directly on AC. Instead, it converts the house current to something in the neighborhood of 12 volts DC. This being so, it is often possible to bypass the power supply and run the equipment directly on 12 volts DC. Anyone with a little electronic knowledge can easily find out what the true DC operating voltage of the device is and find the right place to make the connection.

Obviously it makes sense to use 12 volt equipment wherever possible. For some items however, it is easier to change 12 volts DC to the usual 110 volts AC. Solid state electronic inverters are available that do this job quietly and efficiently. They produce a sort of AC by electronically switching the current rapidly on and off. A step-up transformer raises the artificial AC to the proper voltage. An electric clock powered by such an inverter will run fast or slow, because only the more expensive inverters have precision frequency control. Similarly a record turntable may not run at exactly the proper speed.

While the voltage of house current changes smoothly and gradually from one instant to the next, the output of an inverter changes abruptly and suddenly. This causes many radios and TV sets to produce a loud buzzing sound when connected to an inverter. A final disadvantage of the electronic inverter is that high power units are very expensive.

The obvious way to change DC to AC is to use a DC motor to spin the shaft of an AC generator. This is a motor-generator set, sometimes called a rotary inverter. They are mechanically noisy and less efficient than the electronic inverters, and it is even harder to get them to deliver a constant 60 cycles per second. Fortunately, this is not usually necessary. They produce a smoother waveform, and they are available in sizes that can handle a reasonable amount of power. New ones are expensive, but they can sometimes be obtained as surplus at better prices.

**THE PERFORMANCE AND ECONOMICS OF THE**

# VERTICAL AXIS WIND TURBINE

*Reprinted from Agricultural Engineering February, 1974*

Peter South Raj Rangi
Associate Research Officers
National Research Council Ottawa, Canada

**How much energy is in the wind? What fraction can be extracted and what is this power worth at present rates? Canada's National Research Council (NRC) has some intriguing answers. . .**

The early sail vertical-axis windmills were drag devices—with power produced from drag differential between the arms or sails going upwind and downwind. Thus the machines were limited to a velocity ratio U T/V (rotational tip speed/wind speed) of 0.2 to 0.6 for shielded or hinged paddle wheels and U T/V = 0.5 to 1.5 for a Savonius rotor. The vertical-axis windmill is a simple device. It accepts wind from any direction and the power is available at ground level. However the low velocity ratio and low efficiency made it uneconomical.

The old Dutch windmill with its horizontal axis came around the 12th century. With lift producing blades, the modern horizontal-axis machines are efficient, achieving velocity ratios of 6 to 7. The horizontal-axis windmills, however, have complex rotor, yaw mechanism and drive.

The development of the internal combustion engine provided power on demand and much cheaper than from windmills. Thus interest in wind power has waned.

### Wind Power and its Worth

The potential energy in the wind stream Es of cross section A (sq ft) is $\frac{1}{2}pAV^3$, where p is the density (slugs per cu ft) of the air and V the wind velocity (ft per sec). However the maximum power that can be extracted from the airstream P M, is 16/27 or 59.3 percent of Es, the energy in that stream. The power coefficient of the windmill is

$$C\,p = \frac{\text{power output}}{P\,M}$$

The windmill, due to aerodynamic and mechanical losses, cannot achieve C p = 1. The value of C p for modern

*This is a condensation. For the full report request Paper No. PNW 73-303 from ASAE, 2950 Niles Road, St. Joseph, Mich. 49085. Price is $1.50 per copy.*

Fig. 3 The NRC 12-ft rotor with kinked blades

windmills is about 0.70 —the windmill produces 70 percent of the power that can be ideally extracted.

Here's how the value or worth of this power was calculated. The windmill size and the capacity of the driven device were optimized for average wind speeds ranging from 5 to 30 mph. Data from 14 locations in Canada were used. Then the annual power output was calculated for a given mean wind speed. Assuming different rates for power, the life of the windmills as 10 years and 10 percent interest per annum, what would be the present value of the power that will be available for the next 10 years? If we choose a mean wind speed of 10 mph and the price of power at 1.3 cents per hp-hr, then the present value of that power for the next 10 years is about $2 per sq ft. If we produce a windmill at $2 per sq ft of swept area and have a site with mean wind speed of 10 mph, then the cost of power would be about 1.3 cents per hp-hr.

The price of the horizontal-axis windmills is about $15 per sq ft. At V = 10 mph site the power would cost approximately 9.5 cents per hp-hr. Thus the horizontal-axis windmill could become economical only if the cost could be reduced by a factor of 3 to 5, depending upon the mean winds at the site. This can't be done by increasing the aerodynamic or mechanical efficiency of the windmill. It can only be achieved by a radically simpler design—perhaps sacrificing some efficiency for simplicity.

### The NRC Wind Turbine

Three important factors must be included in a practical windmill design: velocity ratio, design simplicity, and high aerodyanic efficiency. Our first design was a three-bladed vertical-axis rotor with constant chord symmetric aerofoil blades, their span parallel to the axis. When this rotor was given a low rotational speed in an airstream of about 17 fps, it picked up speed and ran at a velocity ratio U T/V greater than 1.

A high-speed rotor with straight, rigid blades parallel to its axis of rotation would be subjected to high bending moment from centrifugal forces and would require extensive bracing. A perfectly flexible blade, on the other hand, under the action of the centrifugal and aerodynamic forces, would conform to a shape in which the only stresses are tensile. The resultant shape of the blade would then be approximately a catenary. Therefore if we curve the blades into the form of a catenary, the bending stresses will be negligible. While the blade wouldn't produce much power near the ends, a constant chord blade would be easy to manufacture.

The optimum rotor configuration would then be the one with constant chord blades curved into a catenary that enclosed the maximum area for a given blade length. For a given length blade curved into a catenary, the swept area is a maximum when the diameter is equal to the height of the rotor.

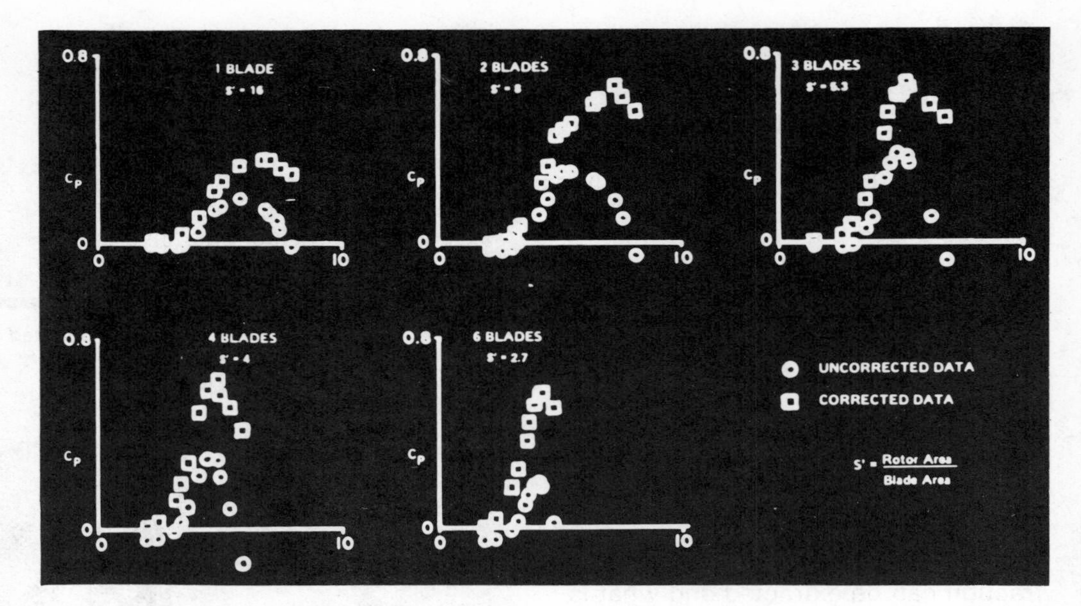


Fig. 2 Power coefficient (C p) - 30-in. diameter rotor

A 30 in. diameter model rotor (Fig. 1) with blades extruded from aluminum of NACA 0012 aerofoils of 1 in. chord was made. The rotor performance was measured in a 6 ft by 3 ft low speed wind tunnel at a constant speed of 17 fps. The solidity of the rotor was varied by changing the number of blades. The rotor power was measured with a dynamometer mounted below the wind tunnel floor.

The test Reynolds number of the blade was only about 20,000. The drag coefficient of the aerofoils is very high at low Reynolds number; therefore, the results were corrected to a full-scale blade Reynolds number of 0.5 by $10^6$. Fig. 2 gives corrected results. The maximum value of C p = 0.70 was achieved with both the two-and three-bladed configurations. With the two-bladed configuration, the maximum C p occurs at U T/V = 7 and with the three-bladed configuration at U T/V = 5.

Because the correction-to-scale effects were large, a full-scale 14 ft diameter rotor was built and tested in the NRC 30 ft by 30 ft V/STOL wind tunnel. The tower and blades rotate in two bearings at each end of the tower. The plate that houses the top bearing is held by guy wires and the lower bearing is housed in a plate fastened to the pedestal. The rotor blades and made from 24-gage coldrolled sheet steel and are approximately NACA 0012.

The rotor output was measured at a wind speed of 18.2 fps with a strain gage torque balance in the two- and three-bladed configurations. The maximum output is 0.67 hp with the two-bladed configuration at about 150 rpm and 0.65 hp for the three-bladed configuration at about 130 rpm.

The output of the rotor was plotted as

Fig. 1 30-in. diameter rotor model

power coefficient C p and is compared with other wind machines. The maximum C p of about 0.65 confirms the corrected results of the 30-in. diameter rotor and compares favorably with C p = 0.7, the efficiency for conventional windmills.

Two 12 ft diameter wind turbines are now installed on the roof of the NAE building. Self-starting devices and simple aerodynamic speed-limiting devices are on the latest models. The blades are now made in two halves. Also, instead of being a continuous curve, they are in three straight sections (Fig. 3) which approximate the original curve. This allows easier manufacture and transportation.

A 15 ft diameter (i.e. 150 sq ft) wind turbine would produce 1 kw at a wind speed of 15 mph; at its probable rated speed of 30 mph, the output would be 8 kw.

**Economics of the Vertical Wind Turbine**

How much will this vertical-axis wind turbine cost? That can't be answered directly until the wind turbine goes into production. We can answer it in directly by comparing it with conventional windmills.

In Fig. 4 the weight per sq ft of swept area for the conventional windmill is plotted against size. The average weight is about 12 lb per sq ft. Conventional windmills cost about 80 cents to $1.25 per lb.

The estimated potential weight per sq ft for the vertical-axis wind turbine, along with the point for the 12-ft prototype model, are in Fig. 4. The estimated weight per sq ft is about 1 lb per sq ft from 100 sq ft to 3000 sq ft swept area. Beyond this size, the weight per unit area is proportional to the square root of the swept area.

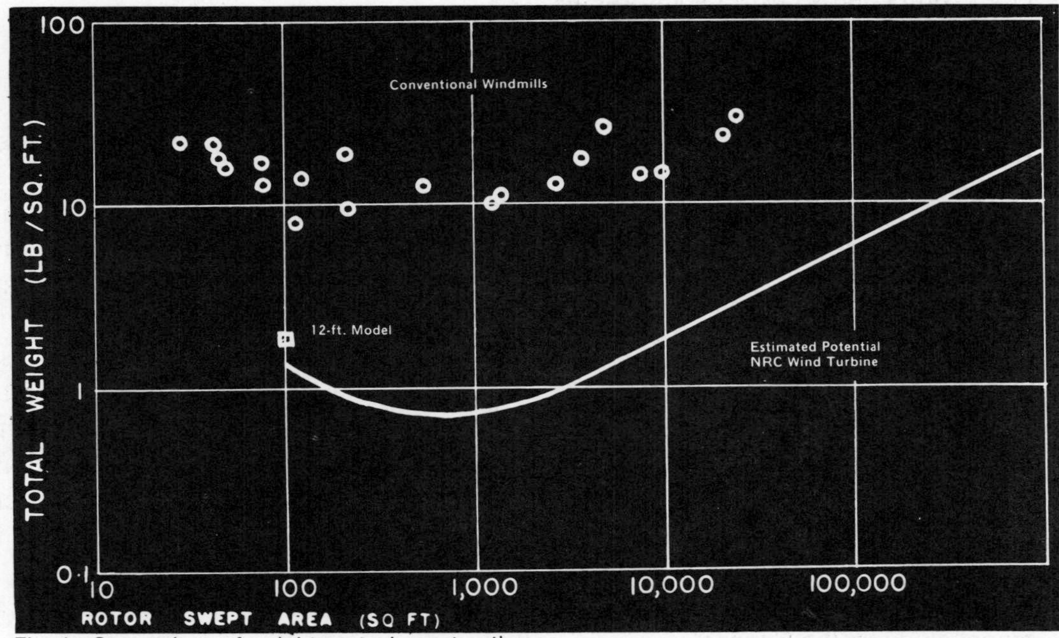


Fig. 4 Comparison of weights—steel construction

Even if we take the weight as 2 lb per sq ft and the cost to be $1 per sq ft, the same as for conventional windmills, then the price for the vertical-axis wind turbine should be about one-sixth that of the conventional windmill ($15 per sq ft). Thus it should be possible to produce the vertical-axis wind turbine for $2 to $3 per sq ft. At these prices the cost of power at 10 mph is 1.3 to 2.0 cents per hp-hr. By producing in large quantities that lower figure of $2 per sq ft can be achieved, which should make windpower available at 1.3 cents per hp-hr. These estimates do not include the cost of storting power for calm periods.

Thus the vertical-axis wind turbine, if produced in quantity, could supplement the conventional sources of power at comparable cost where the annual mean winds are 10 mph or higher.

# Bucket Rotor Wind-Driven Generator

*From Wind Energy Conversion Systems*

By Howard H. Chang
California State University
San Diego, California
and
Horace McCracken
Sunwater Company
San Diego, California

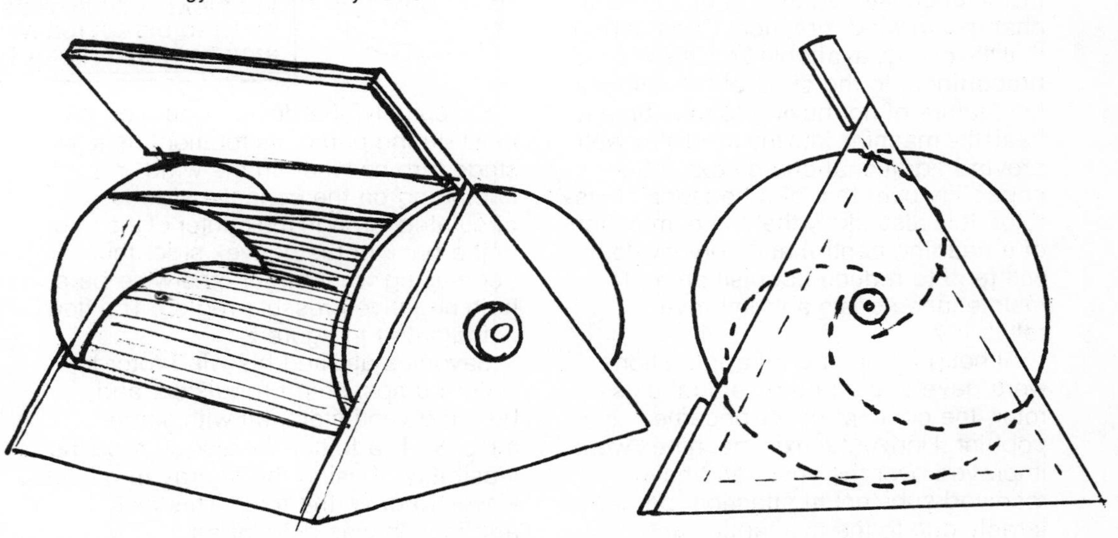

One of the common types of impellers for moving large volumes of air at low velocity is the bucket rotor. Perhaps we have overlooked the possibility of using this design in reverse for the extraction of power from the wind.

To get some preliminary feel for this, a unit with rotor 4 feet in diameter and 4 feet long has been built (see figure 1). It has deflectors on the top and bottom to guide the wind into the top half. The lower deflector also shields the back side of the bucket from the wind, thus reducing the reversing wind force. The present rotor is fixed in direction facing the predominant wind; it may also be mounted and installed with a tail boom to follow the direction of the wind.

Mounted on the trailer towed by an automobile, this unit has been tested at wind speeds from 15 to 40 mph. The mechanical energy produced was measured using a rather crude dynamometer. The maximum power at 40 mph was measured to be 0.14 horsepower. Further improvements in the configuration design will undoubtedly improve the performance.

As compared with the ordinary propeller-type rotor, the bucket rotor is limited in rotational speed since the tip rotor speed can never exceed the wind speed. However, it does not present the blade fatigue problem that the ordinary rotor does, and it perhaps causes less sight pollution. The deflector vanes also provide a venturi passage to capture greater wind flow. The bucket rotors can be strung together end-to-end up to thousands of feet long to produce large amounts of power.

**Discussion**

Q: This is simply a Savonius rotor on its side with a couple of vanes out front to direct the flow. The disadvantage is that it's directional.

A: It is directional, certainly, but because it is directional, the vanes here have the Venturi effect that simply improve the efficiency of the thing.

Q: Right, but, if radial vanes are mounted on a Savonius rotor it captures the wind from all directions, the effect is the same.

A: But the Savonius rotor has a vertical axis. However, if you are free of directional influence, you cannot put in those guide vanes.

Q: You could put them in as part of your supporting structure. They would hold the upper bearing of the vertical axis. I don't see how your rotor is any improvement over the Savonius rotor with guide vanes.

From Wind Energy Conversion Systems

# Vertical Axis Wind Rotors — Status and Potential

By W. Vance
Advanced Concepts Division
Science Applications, Inc.
La Jolla, California

Except for a rather inventive period in the 1920's, the approach taken to extracting power from the wind has been that of using blades or vanes rotating about a horizontal axis with the plane of the blades essentially perpendicular to the wind velocity vector. The two devices shown in figure 1 were patented almost 50 years ago and have as a common element a vertical axis of rotation. The first device, patented by Msr. G. J. M. Darrieus in 1931, has received some recent study by the National Aeronautical Establishment in Canada; the second was developed and patented by Mr. S. J. Savonius in 1929. These rotors share a performance characteristic which differentiates them from the horizontal axis wind rotors, namely, their ability to operate equally well regardless of the direction of the wind. This characteristic is important because it permits the rotor to extract the energy of a given wind or gust instantaneously regardless of any rapid changes in wind direction. Considering that the energy available from the wind is proportional to the cube of the velocity, the feature of not having to take time to head the machine into the wind may well provide additional energy extraction capability over that of a horizontal axis rotor. It is also likely that the elimination of a heading control and servosystem will tend to reduce acquisition and maintenance costs and improve reliability.

Although a number of applications were developed for the vertical axis rotor, the concept never became popular. Horizontal axis machines were improved over the years and have received substantial attention, perhaps largely due to the availability and advance of propeller theory. We believe that the time is right to take a hard look at the vertical axis machines to see if recent advances in aerodynamics, structures, and materials technology might not place these concepts individually (or perhaps in combination) in a favorable light in comparison with the horizontal axis wind rotors.

To maintain brevity, we will concentrate on the S-rotor for the remainder of the presentation. The configuration of the original S-rotor shown in figure 2 resulted from some 30 or more wind tunnel and field tests conducted by Savonius wherein he varied some of the parameters of the rotor.

**Figure 1**

VERTICAL AXIS WIND ROTORS

DARRIEUS ROTOR

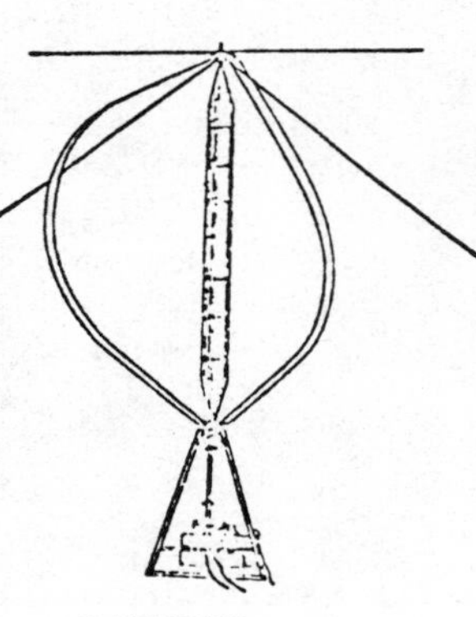

BACKGROUND

PATENTED IN 1931 (US AND FRANCE)

CURRENTLY UNDER STUDY AT NATIONAL AERONAUTICAL ESTABLISHMENT, OTTAWA, CANADA

CHARACTERISTICS

EFFICIENCY ~ 35%

TIP SPEED TO WIND SPEED ~ 6 TO 8

POTENTIALLY LOW CAPITAL COST

CURRENTLY NOT SELF STARTING

S-ROTOR

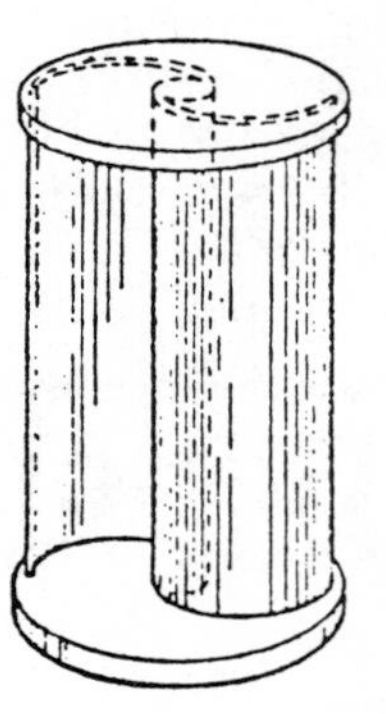

BACKGROUND

PATENTED IN 1929 (US AND FINLAND) BY S. J. SAVONIUS

CURRENTLY USED AS AN OCEAN CURRENT METER

OTHER APPLICATIONS SHOWN FEASIBLE

CHARACTERISTICS

TIP SPEED TO WIND SPEED ~ .8 TO 1.8

EFFICIENCY ~ 31%

SELF STARTING

VERTICAL AXIS ROTORS OPERATE INDEPENDENTLY OF WIND DIRECTION AND THUS HAVE A POTENTIAL FOR HIGH EFFICIENCY IN CHANGING WINDS

Essentially, the device operates (at least during part of its rotation) as a two stage turbine wherein the wind impinging on the concave side is circulated through the center of the rotor to the back of the convex side, thus decreasing what might otherwise be a high negative pressure region. The flow is indicated in figure 2.

Savonius applied his wind rotor to water pumps, ship propulsion, and building ventilators, all with some success. In addition, he also showed the feasibility of using the energy in ocean waves to drive the rotor. This last application was developed subsequentially as an ocean current meter and is available commercially. Very good current measurement capability exists in a region of from 0.05 to 5 knots.

In reviewing the work that has been done on vertical axis rotors, we have concluded that there are a number of development alternatives that should receive some attention from the standpoint of both test and analysis. Figure 3 indicates some of these alternatives. The effects of aspect ratio (the ratio of rotor height to diameter) and the number of vanes will be discussed in detail below. The issue of the profile of the rotor has not been investigated, at least in terms of large (50 ft high or greater) machines. Questions have arisen concerning whether more of the area of the rotor should be at the top to catch the higher wind speeds or whether the area should be at the bottom to provide a more uniform torque distribution along the height. The rotor camber and thickness distribution also need to be optimized. Our own limited amount of test data have indicated that the amount of venting between the rotor vanes has a very significant effect on the rotor speed for a given wind speed.

Figure 4 presents some of the results of a preliminary analysis of the impact of rotor aspect ratio on rotor acceleration. Most of the rotors in use have relatively low aspect ratios (refs. 1 to 3). If we look at the rotor's ability to accelerate as defined by the ratio of the torque on the rotor to its polar inertia, it can be shown that this characteristic improves in proportion to the square root of the aspect ratio as aspect ratio increases. Clearly, there must be limits to this trend due to structural or other considerations. Furthermore, constant-speed performance may impose other requirements.

Test data are shown in figure 5, which indicates the static torque obtained for the two- and three-vaned rotors shown as a function of wind direction. The torque diagram for the two vaned S-rotor has a considerable irregularity that could make it difficult to start under some orientations. The addition of the third vane smoothes the torque diagram to some degree and apparently increases the torque per revolution, but also increases the polar inertia of the rotor, which may offset the increased torque when starting under low wind conditions. Whether two or three vanes will be optimum remains to be resolved. It is also likely that the torque diagram for a rotating rotor may be considerably different from that of the static case described.

The S-rotor may be located in any area where a horizontal axis rotor might be sited. However, the nondirectionality of the S-rotor may be put to use more effectively on sea coasts where the diurnal variation of the wind could be readily accepted. In considering this basic application, it occurred to us that it might be possible to generate an artificial on-shore breeze through the appropriate use of solar energy in the desert. Figure 6 shows a concept of such an artifice. A set of S-rotor are placed circumferentially around a circular area whose surface is made such that the air over it is heated to a higher temperature than the air outside of it. A flow will be established from outside of the heated area to replace the rising heated air. By locating the rotors in the throats of suitably contoured areas, it may be possible to extract considerable energy from the resulting accelerated air. It is recognized that this is an ambitious concept. In essence, we are trying to produce our own wind in sufficient quantities to make a cost-effective power system. Analysis and test techniques must be developed to verify the feasibility of this system concept.

Another application of S-rotor might be in remote areas such as the one depicted in figure 6. In the Arctic and many other places in the world, empty oil drums might be used for rotor vanes. In some underdeveloped countries it may be possible to construct the rotor from indigenous materials. The actual siting of the rotor in a village or base camp would depend on knowing where strong winds persisted without regard to their direction. This simplification coupled with low costs (for the rotor) might make the S-rotor a valuable asset to the community. It should also be noted that vertical axis rotors might be of considerable value in meeting instrumentation and power needs for research on the surface of other planets.

In conclusion, we believe that the potential of vertical axis rotors has not been exploited in recent years and that a comprehensive program including design, analysis, and test could yield devices of equal (if not better) cost-effective performance than that of horizontal axis rotors. We further believe

**Figure 2**

S-ROTOR DEVELOPMENT

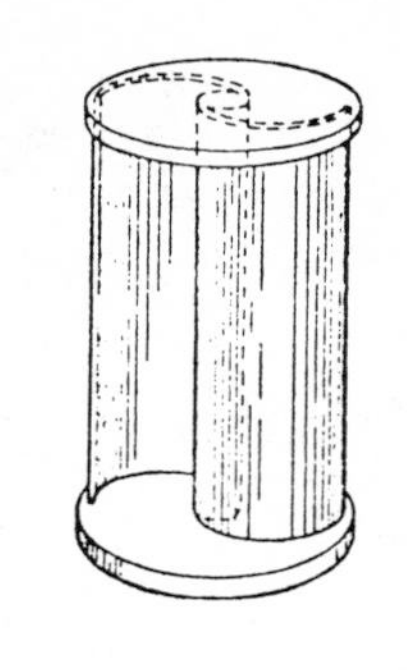

CONFIGURATION

- END PLATES PROVIDE
  - STRUCTURAL MEMBERS
  - CONDITIONS FOR 2 DIMENSIONAL FLOW
- COMMON ASPECT RATIOS < 3
- SHEET METAL VANES

FLOW CONDITIONS

- PERFORMS SOMEWHAT LIKE 2 STAGE TURBINE
- FLOW TO BACK SIDE OF ADVANCING VANE REDUCES NEGATIVE PRESSURE
- SUBSTANTIALLY CONSTANT AREA FOR AIR FLOW

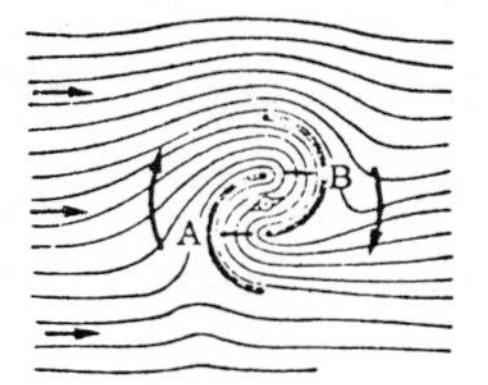


DEMONSTRATED APPLICATIONS

- WIND DRIVEN WATER PUMP
- WIND DRIVEN SHIP PROPULSION
- BUILDING VENTILATORS
- OCEAN WAVE DRIVEN WATER PUMP
- OCEAN CURRENT METER

**Figure 3**

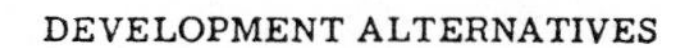


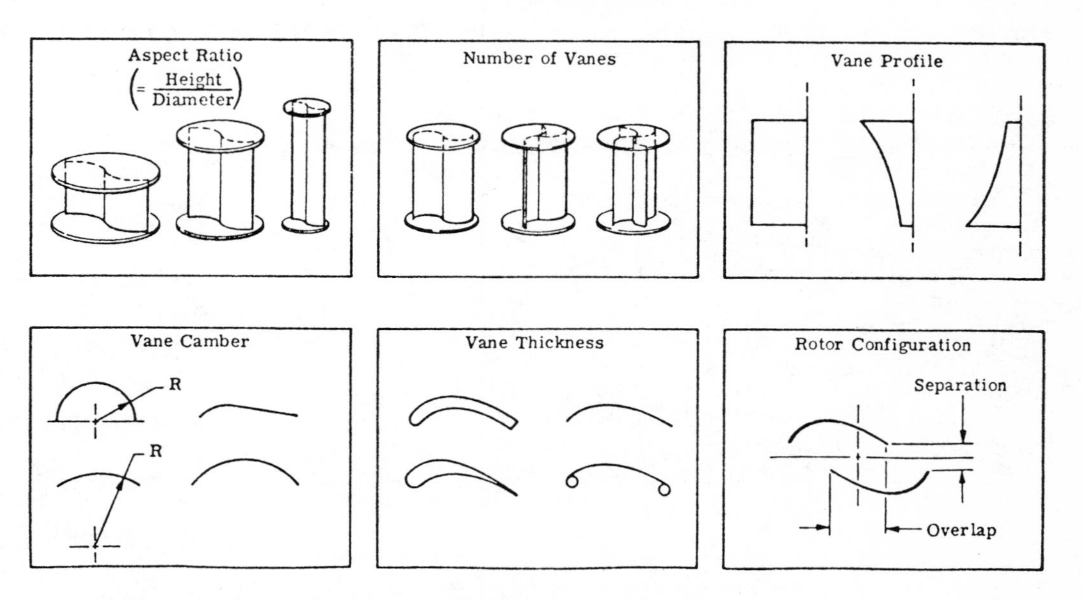


that applications of these rotors should be considered simultaneously with their development to ensure the practical utility of the wind machines.

## Discussion

COMMENT: There are two units currently being manufactured in Switzerland which use this principle. One is a 50 watt unit: the other is a 250 watt unit. The design is a slight variation of the Savonius rotor principle. The large unit has been in production now for about 5 years. They are being used very successfully, particularly on top of radio towers where a little power is needed for a booster amplifier to take the signal down to the house. They worked very well for that purpose.

COMMENT: The company, Electro GMBH Company, also produces the 6-kilowatt standard generators. The man that runs that company is very interested in vertical axis design. He has experimented a lot with them. He has come up with, it seems, a quite successful unit for a small-scale, very simple in design and virtually no maintenance whatsoever, and no problems with regulations in high wind, and so on. There has been one built as a matter of fact, out in the Scripps Institute of Oceanography not too far north of where we are.

Q: You speak of one of these machines being 100 feet high. What is the largest model you know about?

A: I haven't heard of any near that size. An important question is whether this type of rotor be scaled up to large sizes? My answer is: I don't know.

Q: How big you have seen any size.

A: About 15 or 20 feet high.

Q: In the thirties one about 100 feet high was constructed in New Jersey.

**Figure 4** EFFECT OF ASPECT RATIO ON ROTOR ACCELERATION

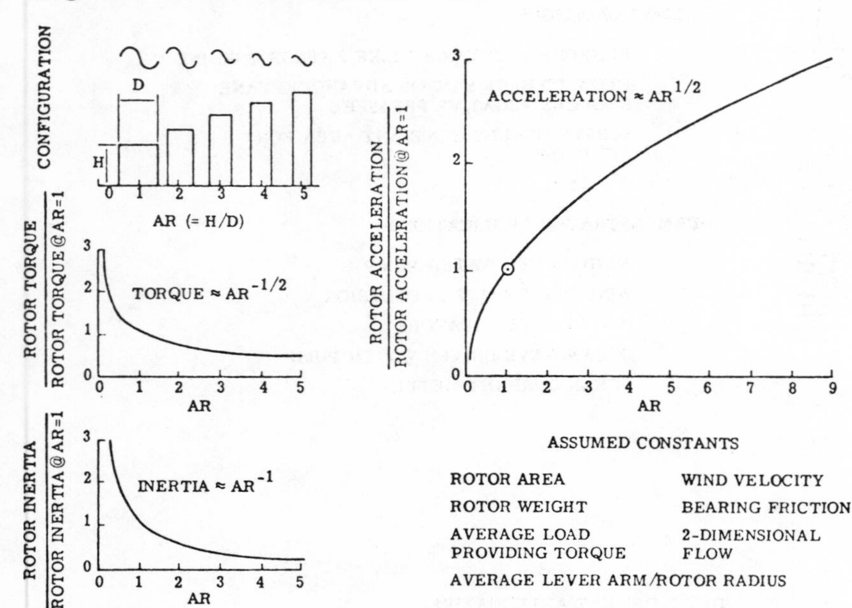


**Figure 5** STATIC TORQUE PROFILE

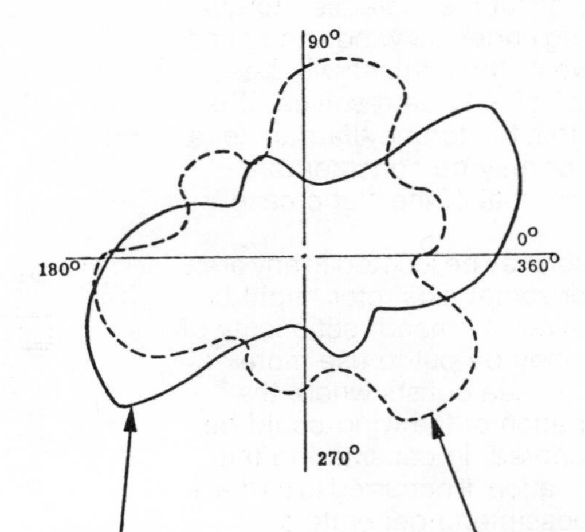


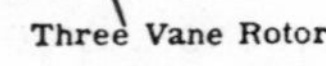


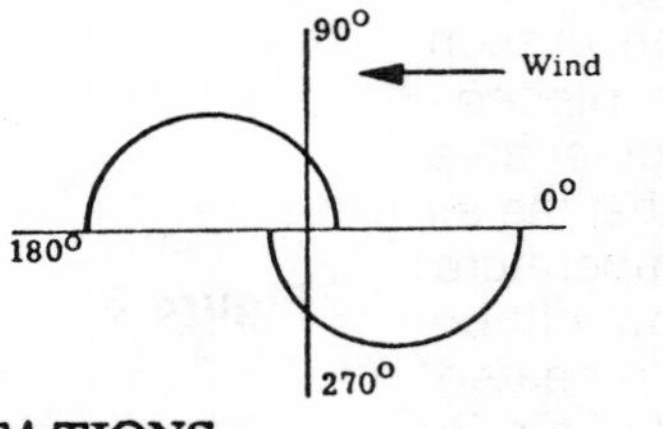


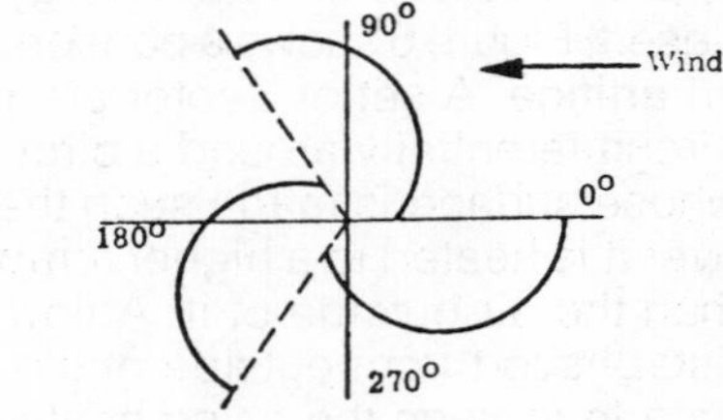


**Figure 6** POTENTIAL IMPLEMENTATIONS

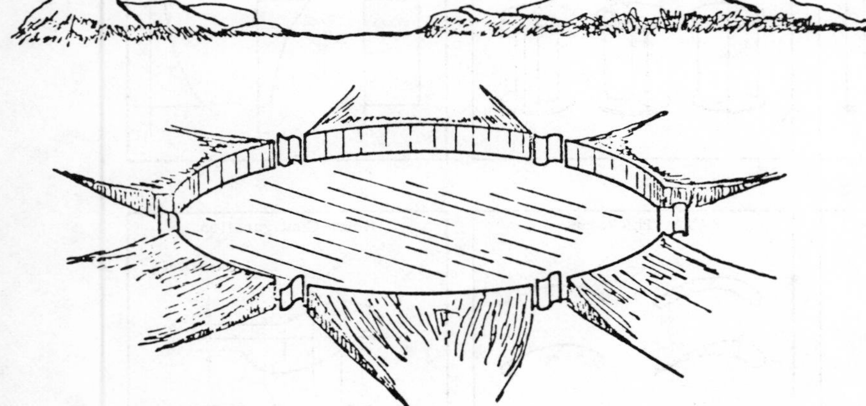

SOLAR/WIND CONCEPT

LARGE RADIATING SURFACE TO HEAT AND RAISE AIR

S-ROTORS ARRANGED TO EXTRACT ENERGY FROM INCOMING REPLACEMENT AIR

FOR EQUILIBRIUM FLOW:

$$V_x \approx \frac{\text{AREA OF RADIATING SURFACE}}{\text{(NUMBER OF ROTORS) (ROTOR DIAMETER)}} V_y$$

POTENTIAL HYDROGEN PRODUCER

REMOTE AREA CONCEPT

SIMPLE VANE REQUIREMENTS PERMIT CONSTRUCTION USING
- OIL DRUMS
- INDIGENOUS MATERIALS

SITING REQUIREMENTS SIMPLIFIED
- DEPENDS ON WIND SPEED -
- NOT DIRECTION

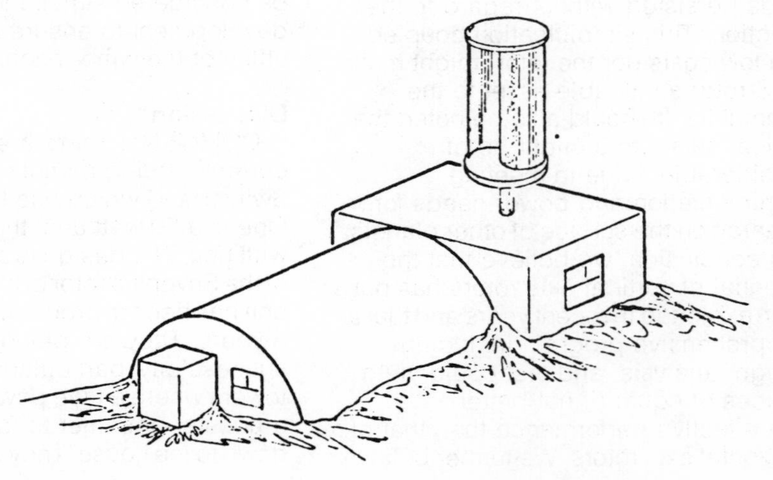

# Power From the S-Rotor

By Michael Hackleman

For generating electricity from the wind's energy, only one type of device has, till now, been easily available: a propeller-driven generator, or alternator. Generally referred to as a Stuart Mill, this unit varies in the number of blades (2, 3, or 4) and the difficulty in shaping the blade's aerodynamic surfaces; the lower end of the price range, however, remains at $400 plus. In this article, an alternative wind device that we have been experimenting with here (at Earthmind) is presented: the Savonius rotor. This unit is low in cost, simple to construct, and boasts some other distinct advantages over the Stuart mill in general performance and safety of operation.

The S-rotor is so called because of its appearance; it is simply a cylinder which has been split into equal halves through its length, had these halves offset by a distance equal to the radius of the original cylinder (approximately), and finally, had these segments secured to end plates which are as large as the new diameter (see drawing A). By inserting a rod through the center of the new assembly, and fixing the rod ends in bearings, the device will rotate when exposed to the wind. If you use coke cans for your original cylinder, you have a toy; but use 55-gallon drums, stacked three on end (and equally out of phase with one another), and there is power in this device at low windspeeds that will surprise you!

How does the S-rotor compare with the Stuart mill in efficiency? If the wind tunnel tests of the two devices were compared, the S-rotor would appear inferior. But under **normal** wind conditions the results are nearly the same! Why? A better understanding of the nature of wind is required to understand this phenomenon. There are two, basic types, of wind: the prevalent, or frequent wind, and the energy wind. The former blows an average of 5 days out of 7, or more frequently than the energy wind, which blows only an average of 2 days out of 7; this means that 70% of the energy comes from winds blowing only 40% of the time. Energy winds come, mainly, in the form of gusts, and it is a characteristic of gusts that they usually deviate in direction from a prevalent wind by 15-70 degrees. So what?

The significance of this can be demonstrated by placing a Stuart mill alongside a S-rotor in a steady wind. Suddenly, there is a gust. The Stuart mill swings in the gust. The gust dies. The tail of the Stuart mill slowly swings it back into the steady wind. The S-rotor,

meanwhile, sped up with the gust and slowed as it died. The Stuart mill lost much of the energy from both the gust and the steady wind, because it took time for it to align itself with the two winds. The S-rotor, however, was able to absorb the full energy in both the gust and the steady wind because it didn't have to "swing" (or track); one of the greatest assets of the S-rotor, then, is that it can take a wind from **any** direction at **any** time.

Incidently, in a steady, low speed wind the swinging of the Stuart mill is acceptable but at higher speeds it is not. The spinning propeller is just one big gyroscope, and the constant adjustment to wind direction exerts tremendous forces; the result is gyroscopic vibration and it is responsible for sending many a propeller, generator, and tower crashing to earth! Because it does not have to swing, the S-rotor experiences no such problem!

Another thing becomes apparent when watching these two wind devices side by side; the S-rotor appears to turn very slowly, for the blades of the Stuart mill may rotate as many as eight times for every rotation of the S-rotor. But if you think that speed is a necessary ingredient to producing power, think again. The Stuart mill has to obtain high rotational speed to get this power; the S-rotor, however, presents many times more surface area to the wind, and develops this same power at low rotational speeds. Higher gear ratios are required for the S-rotor (to get the alternator up to the necessary speed for charging) but it may use them without experiencing re-start problems, a factor which prevents the Stuart mill from using them. Additionally, the S-rotor, with its low rotational speed, requires minimal or no balancing, and simple tools are used to construct its wings. Because of the high rotational speeds achieved by the Stuart mill, the blades must be well designed and balanced; as few folks are equipped with the tools or the talent to do this, the blades (or the whole propeller) must often be purchased, and at higher costs to the buyer.

What else is evident? Well, the Stuart mill rotates horizontally and is mounted, along with its generator, on top of a tower. The S-rotor? It rotates about a vertical shaft, which means that its alternator can be mounted on, or near, the ground. This also means—no tower! Just a pole, with some guy wires. Easy access to the alternator, easy lowering of the S-rotor, and easy relocation. **Minus** the expense of the tower!

Preliminary tests of the Earthmind prototype indicates the S-rotor can begin charging 12 volt batteries at wind speeds lower than the 7 mph minimum required for the Stuart mill; we've done it at 6 mph. We will continue research into this area (lower charging windspeeds)

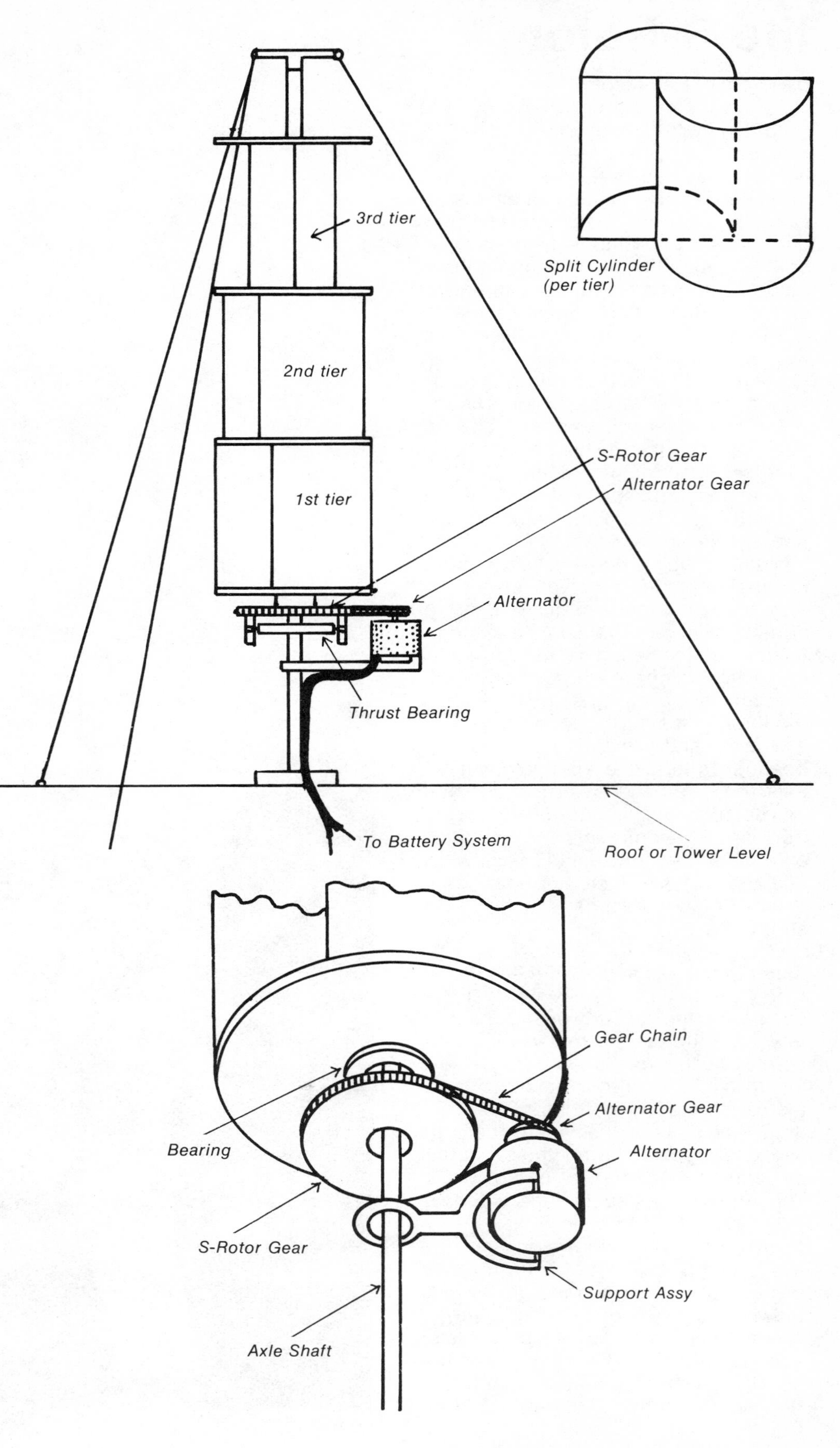


*Prototype S-Rotor* *Original Design*

as it would greatly benefit people living in areas of low, average wind speeds that might not otherwise be able to install a conventional wind energy system of the Stuart mill type.

Construction of the Savonius unit includes a centrifugally-operated spoiler for slowing the rotor in extremely high windspeeds, and a separate sensor (also mechanical) which is used to limit, or switch off, field current to the alternator for high wind, or no wind, conditions. The S-rotor uses the same alternator and, if required, a similar inverter to those used in the Stuart mill.

An additional sensor is added to our Earthmind prototype for the operation of a data collection system, which will supply us with the information for evaluating S-rotor performance and support the claims that we make for the device. This system operates automatically anytime that the windspeed goes above 3 mph and records data from a separate wind speed and direction indicator as well.

As the S-rotor can be built from more readily accessible materials, the cost of construction will almost always be less than $100 (not including the cost of batteries or an inverter). We intentionally purchased all the parts that went into construction our prototype and spent $103. This includes the alternator, gears and chain, bearings, pipe, stock rod, screws, bolts, eyebolts, guy wire, turnbuckles, paint and stain, and misc. springs, wire, etc. Our second, and larger, S-rotor will cost half of that amount as we have obtained most of the parts without expense (55-gallon drums, etc.); we have everything but the bearings, gears and chain, and the gas for the welding unit. The first S-rotor did not require any welding; the second unit (under construction) will be welded together from convenience and the time that it will save us (but it is not necessary).

A lot of thought should go into the types of batteries to be used in this, or any other, wind energy system. If money is not a consideration, then NiCads (Nickel-Cadmiums) are good and I would certainly recommend them; initially, however, we will use lead-acid batteries because of a low cost, special "deal" that we have been able to work out with a local auto wrecking yard (car batteries, like trash, are a growing resource). For those folks that might be able to duplicate what we've worked out, here is what we did. We gave the guy who runs the yard a $10 deposit (this is his insurance that he won't get ripped off) and his job is to inspect incoming batteries of equal, or greater than, 100 AMP hour capacity, and to give us a call when he has one. We pick it up and keep it for a week of testing; we have a hydrometer, cell voltage tester, a charger, and a dummy load (to

discharge the batteries at a known rate) and I've used all my own knowledge on batteries to set up a program of rigorous testing. If the battery checks good, we pay the yard man $2; if it tests poorly, we return it and the yard man disposes of it through his regular outlet (where he can only get $1.25 in this instance). As we only need **ten** of the 12 volt batteries, we get them all for $20 plus the $10 goes to the yard man if he can get all of the units within a predetermined period of time (we get all of the deposit back if he doesn't). He can't lose and often gains (a maximum of $17.50 **above** his normal income for the batteries. We always get a good battery, with no risk, at low cost. Proper care and use and a lead-acid battery (that is in good condition) results in its lasting many times what the same battery will last when used in an automobile and exposed to its conditions. This goes for all batteries; even Ni Cads won't last long if they aren't properly charged, discharged, and maintenanced.

Incidently, the Savonius rotor has a long history of application in pumping water; we are particularly pleased with a unit designed by the Brace Research Institute work in Canada. Detailed plans, for example, for the construction of an S-rotor water pumping unit that costs $51.00 to build are available from them for 75¢. They may be obtained by asking for leaflet L-5 and writing them at: Brace Research Institute, MacDonald College of McGill University, Ste. Anne de Bellevue 800, Quebec, Canada. Ask them for a list of their publications, too, as they have some really nice information on some other wind and solar energy projects they've done, or are doing.

*Earthmind is a non-profit research and educational corporation. They are doing research into alternative sources of energy, better nutrition, ecological farming, and gardening, and other activities intended to transform and humanize our culture. A stamped self addressed envelope will bring you a list of their publications.*

*More detailed information on the design, construction, and performance of the Savonius rotor is available in a book called "Wind and Windspinners" put together by Earthmind. It costs $7.50 and you can order it from Earthmind, Josel Drive, Saugus, Ca. 91350. Earthmind appreciates feedback. If you build an S-rotor after ordering the book, they will give you a $3 refund for photos or information about your plant.*

# Savonius Rotors

*From Alternative Sources of Energy*

By Ted Ledger

Fifty years ago, I made a very simple mill from a 50-gallon oil drum. The drum was split axially, spread out and welded to a Model-T Ford torque tube. This gave a pair of anti-friction bearings for the split drums to turn on a vertical axis. (Fig. 1.)

The vertical axis makes it unnecessary to have any tail or rotating system. It also allows the rotary motion to be brought down the support, or the tower to the ground if necessary.

The mill I made ran a rotary water pump for at least 30 years and was finally destroyed by flooding of the lake when a company was permitted to raise the level for power purposes.

The overlap of the drums increases the efficiency of the mill as the air from the driving drum blows off the turbulence behind the idle drum. (Fig. 2.)

As a test, make a model out of a split 48-oz juice can. Use gas tubing for a sleeve-bearing and solder it together. Both the model and the full-scale one need some balancing to prevent vibration at speed. To do this, suspend the rotator and let the wind turn it. It will find its dynamic center and this can be marked on the underside while it's turning. Load the bottom to correct the error. These two corrections interact a little, so it may require a little adjustment. Use putty or plasticine for temporary balancing loads. Replace with permanent weights when the job is satisfactory.

I don't know who made this mill first, I didn't. I saw a picture a long time ago and built the water pump I mentioned.

Today, Model-T Ford torque tubes are museum pieces, but a welder can come up with a good substitute with car bearings and tubing.

The drums have stiffener-sheets welded in at the point shown, to prevent the edge from flapping.

The cut edge could be hemmed or improved with a slip-cover.

If more information is needed, write to:

Ted Ledger
591 Windermere Avenue
Toronto, Canada M6S 3L9

**Figure 1 - Top View**

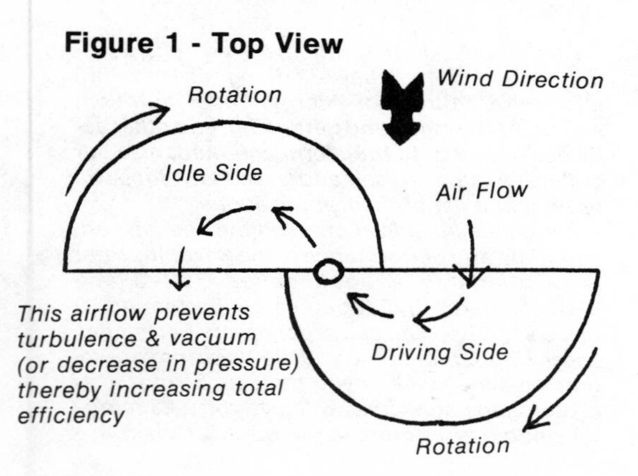


**Figure 2**

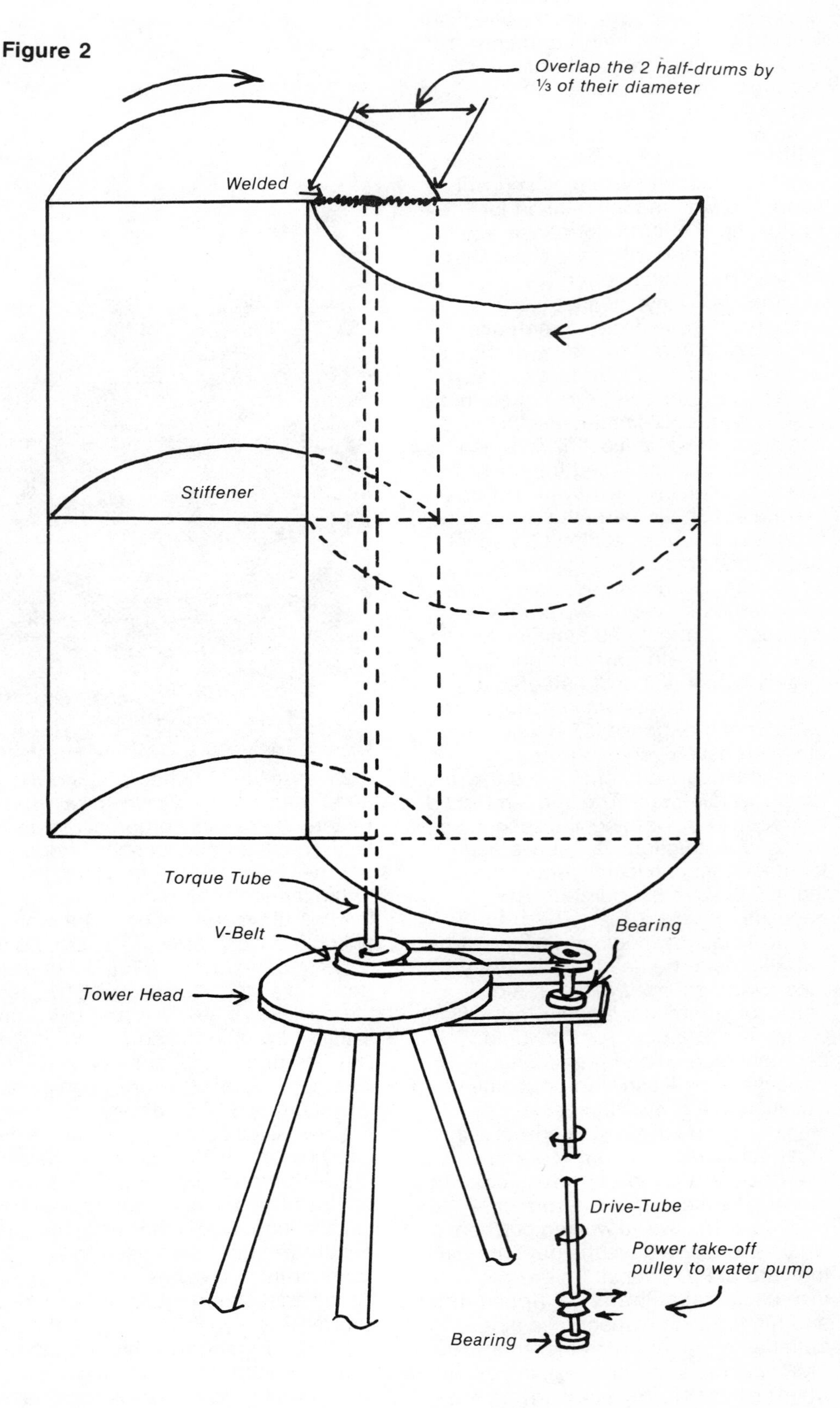

# Methane

Methane composes about 95% of natural gas. It burns cleanly and efficiently and is one of our most popular fuels. Because of its popularity, it is being rapidly used up, and some experts predict that most of our natural gas will be gone in about thirty years. The situation has been made worse by government controls that kept prices down and encouraged consumption, while discouraging the search for new reserves. Now that a scarcity is likely, there is a scramble to develop new sources of this useful gas.

Methane is a fossil fuel, found in oil wells and coal mines. However, it is also produced by the decay of organic matter in the absence of oxygen. Methane is found in swamps, septic tanks, and the digestive systems of animals. In this respect, methane is different from oil and coal. Once these are used up, they are gone forever, but it is easy to manufacture new supplies of methane.

When found in coal mines, methane is known as "fire damp." Many mining disasters have happened when a spark ignited an explosive mixture of air and methane. In modern times, ventilating systems sweep the gas out of the mines. However, the Department of the Interior has a scheme to put the gas to use. In Bula, West Virginia, methane is being drained from a coal seam and sold commercially. The idea is to drill holes into new coal seams and tap the gas for two years before mining starts. If successful, it is thought the practice could double our natural gas reserves.

Methane is also produced by vegetation decaying underwater. It bubbles to the surface as "swamp gas." When ignited, it produces the dancing swamp lights known as "will-o-the-wisp" and "fool's fire." Methane can also be produced by buried waste. This is the idea behind an attempt to recover methane gas from a landfill at Palos Verde, near Los Angeles. Thirteen million tons of refuse have been placed in the landfill since 1957. The contractor hopes to get 1,000 cubic feet of gas per minute. If the project is a success, there are about a thousand other landfills in the U.S. that could be similarly treated.

Another idea is to produce a natural gas replacement by roasting coal. This was done in many cities before WW II, but was dropped when supplies of inexpensive natural gas became available. Now modern technology is being applied to the problem. Coal is plentiful, but the process has its disadvantages. The fuel gas produced has only about half the heating value of natural gas and requires further processing to improve its fuel value. The process uses large amounts of strip mined coal and large amounts of water, which is a problem if the gas is to be produced in the arid West where the coal is mined.

Many people are becoming excited by the idea of producing their own methane. This is done by allowing manure and other organic wastes to decompose in a sealed tank called a digester. Bacteria break down the organic matter, producing a mixture of methane and carbon dioxide called bio-gas. Bio-gas contains from 50-70% methane. Most of the rest is carbon dioxide, with traces of nitrogen, hydrogen, and hydrogen sulfide. This can be used for the same purposes as natural gas. As a bonus, the sludge left in the digester when it has stopped producing gas makes a fine fertilizer.

Many sewage plants have used the process since the 20's to reduce the volume of their sludge. The gas produced is sometimes used in the operation of the plant, but often it is simply burned to get rid of it. During WW II farmers in fuel-starved Europe built their own gas plants, and some even ran tractors with the gas. Today most of the research in this area is being done in India. There dung which could be used to fertilize the fields is burned for fuel, because other fuel is extremely scarce. The use of digesters could break this cycle by producing bio-gas for fuel and a rich sludge for use as fertilizer.

In his novel **Needle** science fiction writer Hal Clement portrays a future in which whole Pacific Islands are devoted to tank farms where specially tailored bacteria are used to break down jungle vegetation, producing not only methane but synthetic petroleum as well.
This does not seem too likely right now. If all the manure and garbage we can conveniently collect were digested to produce methane, it would satisfy only about 7% of our demand for natural gas. Clearly the methane digester is not a total solution to the energy shortage. Nor is it a practical proposition for a single family. The wastes produced by one family would not generate enough gas to supply their needs. A farm, however, is an ideal place for a methane digester. The digester supplies not only fuel and fertilizer, but a convenient means for disposal of manure. With two billion tons of manure being produced on U.S. farms and feedlots each year, that is no small benefit. The waste problems of the cities are growing even greater too, and perhaps someday soon waste disposal problems plus rising gas costs will make it practical for cities to produce and process bio-gas to supplement their gas supplies.

The process that goes on inside a digester takes place in several stages. When the digest is first loaded, bacteria begin to break down the organic matter, turning it into water and carbon dioxide. They need oxygen to do this, so they are called **aerobic** bacteria. When the oxygen inside the digester is used up, a different set of bacteria takes over. These do not need oxygen. In fact, it kills them. Because they need oxygen-free conditions, they are called **anaerobic** bacteria. Without oxygen to work with they are unable to break down organic matter as completely as the aerobic bacteria. This is why the process of anerobic fermentation yields methane.

The anerobic fermentation is a two step process. In the first step, a group of bacteria breaks down complex organic molecules into simpler ones. Some of the simpler molecules are organic acids, so this group of bacteria is sometimes called the acid formers. They are the ones responsible for turning cider into vinegar. When a stock of simple molecules has been built up, another set of bacteria goes to work. They break the simple organic molecules down further, and they are the ones that do the actual job of producing methane. Up until this point, the digester has been producing nothing but incombustible carbon dioxide. Its contents have been slightly acid (Ph about 6) because of the work of the acid forming bacteria. When the methane formers start working, the mixture turns somewhat alkaline (Ph about 7.5 to 8.5). The methane producers are more delicate than the acid formers and must be pampered if they are to do their job well. Building a small batch fed digester is a very simple project. No high pressures or temperatures are involved, so the digester can be almost any airtight container. Size and shape do not matter, and digesters have been made from inner tubes, oil drums, pressure tanks, etc. The larger the digester, the more manure it will hold, and the more gas it will produce. One pound of fresh cow manure will generate roughly one cubic foot of gas. A cubic foot of manure will generate roughly 50-60 cubic feet of gas. Since the manure will be mixed half and half with water before being loaded into the digester, about 25-30 cubic feet of gas will be generated for each cubic foot of digester capacity. Consider the digester's maximum capacity to be about 85% of its volume. Some space must be left at the top in case of foaming.

Garbage, paper, grass clippings, in fact almost any organic material can be fed to a digester and used to produce bio-gas. In general, however, the best results are obtained when the mixture is mostly manure. The digestion process is not critical, but if care is taken to provide the bacteria with favorable conditions, the gas produced will have a better methane content and a higher fuel value.

According to Ram Bux Singh, leading Indian methane researcher, digestion proceeds most rapidly and yields the best quality gas when the material in the digester has a carbon to nitrogen ratio of about 30 to 1. Singh and others have published charts showing the C/N ratios of various materials, but these can serve only as rough guides. The figures vary a great deal because the C/N ratio depends on such things as age, diet, growing conditions, and so forth. Cow and horse manures are best for direct digestion, with C/N ratios of about 20. Other manures have lowr C/N ratios. Additional carbon in the form of plant wastes can be added to raise the ratio, but Ram Bux Singh suggests adding no more than one part vegetable waste to one part dry manure. (Or about one part vegetable waste to six parts fresh manure.) If the C/N ratio is too high, digestion slows down. There is another reason for limiting the amount of plant material in the digester. Not all the carbon contained in plant wastes is available to the bacteria. Some of it is in the form of lignin, a component of cellulose which the bacteria cannot digest. The use of too much rough plant material can lead to the formation of scum which clogs the digester.

When feeding a digester, the usual practice is to mix the material with enough water to make a mixture or slurry with the consistency of thick cream. Plant wastes should be chopped or shredded if possible. Mixing the material with water makes it easier to load, and it disperses the material allowing better contact with the bacteria. For best results the slurry should be from 7 to 9% solids. Ordinary cow manure falls in this range when mixed half and half with water. If the slurry is too thin, much of it will settle to the bottom of the digester and digestion will be slowed.

Ideally, the temperature of a digester should be kept between 80 and 105°F. The methane producing bacteria work best in this range. Activity will continue at lower temperatures, but methane is produced more rapidly and in greater quantity if the temperature can be kept at the right level. In cold climates, digesters must be insulated and supplied with heat. Often part of the digester's output is burned to keep it warm. This may use 20-30% of the gas, but the increased production makes it worthwhile. Some have suggested the use of solar collectors to heat digesters. This would allow the replacement of a high quality fuel (methane) by a low quality heat source (solar heat) and would be a much more efficient use of energy.

If left to itself the material in a digester would settle into layers, with solids on the bottom and liquid on top. For most efficient digestion, the contents should be agitated from time to time. This keeps all the material in close contact with the bacteria. Frequent agitation also helps keep scum from building up. Scum is made up of the indigestible material that floats to the surface. If allowed to build up, it can form a dense mat that clogs the digester completely. Agitation breaks up the scum layer and helps keep it from interfering with digester operation.

After 30-40 days a digester will have produced most of its methane. At the end of this time it can be opened, emptied of spent liquid and sludge, and refilled with a new batch. The problem with operating a digester in this way is that its production is not uniform. A great deal of gas is produced in the beginning and then production gradually trails off. One way to overcome this is to operate several digesters on a staggered schedule. Another way is to feed the digester continuously.

After the digester has settled down to regular gas production, a small amount of new material is added daily and a corresponding amount of sludge is withdrawn. If the material being used takes 40 days to digest, then the amount should be 1/40 of the digester's capacity. In order to prevent material from being withdrawn before it has had a chance to be digested, vertical digesters are equipped with baffles. Long horizontal cylinders also make good continuous digesters. They resemble, in shape as well as in function, the intestines of an animal. Continuous feed digesters produce gas more steadily and only small amounts of manure and sludge need be handled at any one time. They must be fed with care, however, for too much fresh material can upset conditions in the digester. If this happens, feeding should be stopped until the digester has a chance to get over its "indigestion." Eventually continuous digesters must be shut down to remove scum. In this respect, horizontal digesters have an advantage over vertical ones, since they have more surface area and it takes scum longer to build up.

The sludge from a digester is an excellent fertilizer. It contains more nitrogen than ordinary compost. The digestion process does not destroy all disease producing organisms, so as a matter of common sense, the sludge should not be used on any vegetables that will be eaten raw.

Bio-gas can be used like natural gas for cooking, heating, and lighting. It can also be compressed for use in internal combustion engines. Unfortunately, it takes a great deal of work to compress the gas, so it is not an efficient use of energy.

Because the methane in bio-gas is diluted by unburnable carbon dioxide, it has a lower fuel value than natural gas and more of it must be burned to do the same job. Burner nozzles should be about 30% wider and fuel lines should be correspondingly larger. Because bio-gas contains water vapor, condensation traps should be included in the system to keep water out of the gas lines.

Bio-gas can be used safely if it is treated with the same respect given to natural gas. A mixture of 5-14% bio-gas with air is explosive. For this reason, the first batch of gas from a digester should be vented without trying to ignite it. This flushes the remaining air out of the system. This should also be done if there is any suspicion that air has entered the system. It's also a good idea to bubble the gas through a flame trap before it is used. If by some mischance there is an explosive mixture in the system, the flame trap prevents the flame from reaching the digester and the gas storage.

Bio-gas generation is inherently a simple and inexpensive process, especially when compared with some of the energy sources now proposed. Perhaps one of its greatest appeals is that it is well within the reach of the individual. Indeed, it is possible that the backyard tinkerer can make significant contributions, by finding ways to build better digesters or methods for improving gas yield and quality.

# Simplest Methane Experiment Yet

No special equipment is needed to generate methane. In fact, you've been doing it all your life without realizing it. The process takes place in your lower intestine, where it's dark and damp and oxygen-free. These conditions are ideal for anaerobic bacteria, which flourish on the residues of the digestive process. They produce hydrogen, hydrogen sulfide, and methane. Certain combinations of food are better at producing gas than others. Beans are famous in this respect. There is usually from 50 to 150 milliliters of gas in the lower intestine and up to ten percent of this is methane. This methane is just as inflammable as methane produced in other ways. Collecting and igniting the gas is a favorite amusement of medical students. Others, who scorn the use of apparatus altogether, simply bend over and have a friend hold up a lighted match at the proper moment. The effect is most spectacular when the experiment is performed in a dim light.

# Enzymatic Hydrolysis of Cellulosic Wastes

By M. Mandels, J. Nystrom, L. A. Spano
U.S. Army Natick Laboratories
Natick, MA 01760

Cellulose is our most abundant organic material which can be used as fuel. The net world wide production of cellulose is estimated at one hundred billion tons per year. This is approximately 150 lbs. of cellulose per day for each and every one of the earth's 3.7 billion people. The energy to produce this vast quantity of cellulose comes from the sun and is fixed by photosynthesis as discussed by Dr. Brown. The energy from the sun, available over the United States alone is between 4 and 5 x $10^{19}$ BTU/Yr.

This is aproximately 600 times the annual energy consumption of the United States. Prior to 1900 our principal sources of energy were the wind, wood, water power and coal. During this century we have been relying very heavily on fossil fuels originally produced by photosynthesis. Our energy consumption in the United States has been estimated at 7 to 8 x $10^{16}$ BTU/Yr. This total energy is obtained primarily from oil (43%), gas (35%), and coal (19%) (1). Comparison of the annual energy consumption in 1873 (4.2 x $10^{15}$ BTU/Yr) with that of today, shows that our current demand is approximately seventeen to twenty times more than what we used in 1873. This phenomenal growth in energy demand will be difficult if not impossible to support with our current fuel reserves regardless of processing capabilities.

By year 2000, undoubtedly nuclear power may be a major source of energy; however, to achieve the ultimate goal of independence, we will have to harness effectively and economically the inexhaustible energy of the sun.

Since cellulose is the only organic material that is annually replenishable in very large quantities, we should explore many ways to utilize it as a source of energy, food, or chemicals. The utilization of this resource is greatly simplified if cellulose is first hydrolyzed to its **monomer** glucose as shown in Figure 1. Once we have formed the glucose, it can be used as a food consumable by man and animals, it can be converted to chemical materials, it can be converted microbially into single cell proteins, and it can be fermented to clean burning fuel (ethanol), solvents, (acetone), other chemicals, etc. It is estimated that from one ton of wastepaper we can produce ½ ton of glucose which can be fermented to produce 68 gallons of ethanol. Several

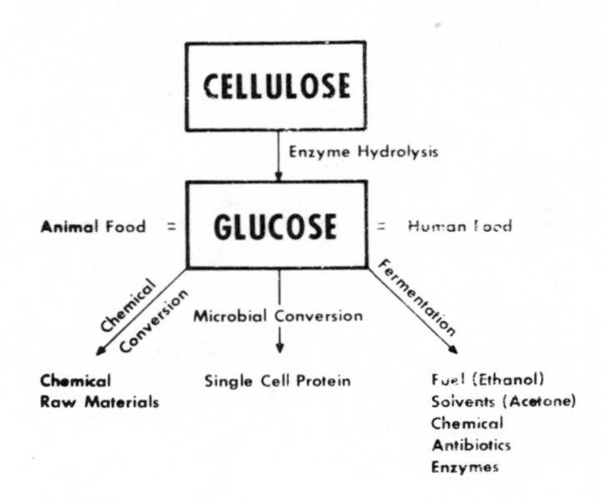


studies conducted in the past several years to determine the suitability of blending methanol and ethanol with gasoline for use in internal combustion engines, have shown that this can be done easily with only minor modifications if any to present engines. Moreover, it has been found that engines burning these blended fuels have fewer problems with exhaust emission (2), (3).

The simplicity of hydrolyzing cellulose to glucose and converting the latter to chemical feedstocks to conserve petroleum which is now used to make petrochemicals or fermenting the glucose to ethanol that can be easily blended with gasoline to power automobiles and other internal combustion engines, could alleviate our immediate energy crisis.

The shortage of fuel that has been estimated at 2.5 to 5.0 million barrels per day could be easily met by the daily hydrolysis of 1.5 to 3.0 million tons of waste cellulose present in municipal trash and agricultural wastes. Conversion of cellulose to glucose can be done by either acid hydrolysis or by enzymatic processes. (4-14). There are various advantages in the use of enzymes to hydrolyze cellulose instead of acid. When using acid, expensive corrosion proof equipment is required. Waste cellulose invariably contains impurities which will react with the acid producing many unwanted by-products and reversion compounds in the digest. The enzyme on the other hand is specific for cellulose so that the glucose formed is fairly pure and constant in composition.

We at the U.S. Army Natick Laboratories are developing an enzymatic process, which is based on the use of the cellulase enzyme derived from a mutant of the fungus **Trichoderma viride** isolated and developed at the Natick Laboratories. A schematic diagram of such a process is shown in Figure 2. Our first step is the production of the enzyme. This we accomplish by growing the fungus **Trichoderma viride** in a culture medium containing shredded cellulose and various other nutrients. After 5-10 days the fungus culture is filtered and the solids discarded. The clear straw colored filtrate is the enzyme solution that is used in the saccharification reactor. Prior to its introduction into the reactor, the enzyme broth is assayed for cellulase and its acidity adjusted to a pH of 4.8 by addition of a citrate buffer. Milled cellulose is then introduced into the enzyme solution and allowed to react with the cellulase to produce glucose sugar. You will note that saccharification takes place at atmospheric pressure and low temperature 50°C. The unreacted cellulose and enzyme is recycled back into the reactor, and the crude glucose syrup is filtered for use in chemical, microbial, and/or fermentation processes to produce chemical feedstocks, single cell proteins, fuels, solvents, etc.

**Conversion of Waste Paper Products to Glucose Sugar**

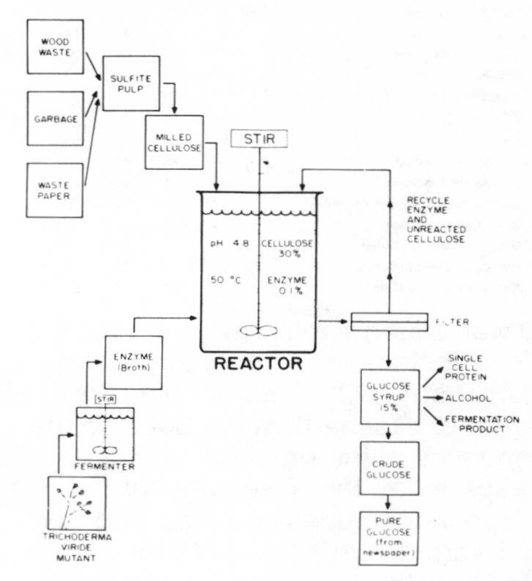


The key to this process is the production of high quality cellulase from **Trichoderma viride**. During the past twenty years, extensive studies of this fungus and its enzyme have been made at the Natick Laboratories in connection with the program on prevention of deterioration of cellulosic materials. For this process, today we are interested in accelerating deterioration. To date we have defined the conditions needed to produce the enzyme in quantity. We have also developed mutant strains that produce 2 to 4 times as much cellulase as the wild strain. In this area we feel that we have yet to reach the upper limit.

Another important variable to be optimized is the preparation of the substrate. The insolubility and crystallinity of pure cellulose and the presence of lignin in waste cellulose make it a most resistant substrate. The most satisfactory pretreatment we have found is ball milling. This reduces the crystallinity and particle size of the cellulose and increases its bulk density. Consequently more cellulose is available for saccharification in the reactor. Figure 3 shows the hydrolysis of a number of pure and waste celluloses by the culture filtrate of **Trichoderma viride**. Saccharification is slow for crystalline cellulose such as cotton or untreated rice hulls or bagasse. Pot milling greatly increases their reactivity. Shredded or milled papers make good substrates. The Black-Clawson fiber fraction from the hydropulping separation of municipal trash, is an excellent material especially after milling. The same is true for the cellulose fraction separated by dry air classification of municipal trash by the Bureau of Mines' process. Since these waste materials contain impurities, hydrolysis is limited to the cellulosic fraction of the substrate.

HYDROLYSIS OF CELLULOSE BY TRICHODERMA VIRIDE CELLULASE

| Substrate | % SACCHARIFICATION 1 hr | 4 hr | 24 hr | 48 hr |
|---|---|---|---|---|
| PURE CELLULOSE | | | | |
| Cotton – Fibrous | 1 | 2 | 6 | 10 |
| Cotton – Pot Milled | 14 | 26 | 49 | 55 |
| Cellulose Pulp SW40 | 5 | 13 | 26 | 37 |
| Milled Pulp Sweco 270 | 23 | 44 | 74 | 92 |
| WASTE CELLULOSE | | | | |
| Bagasse | 1 | 3 | 6 | 6 |
| Bagasse - Pot Milled | 14 | 29 | 42 | 48 |
| Corrugated Fibreboard Mighty Mac | 11 | 27 | 43 | 55 |
| Corrugated Fibreboard Pot Milled | 17 | 38 | 66 | 78 |
| Black Clawson Fibers | 5 | 11 | 32 | 36 |
| Black Clawson Pot Milled | 13 | 28 | 53 | 56 |
| Bureau of Mines Cellulose | 7 | 16 | 25 | 30 |
| Bureau Mines Pot Milled | 13 | 31 | 43 | 57 |

It was stated earlier that pretreatment of the substrate is an important variable. This variable will affect not only the degree of saccharification but also the economics of the process. Using newspaper as the base substrate, various techniques were tried and the results are shown in Figure 4. It should be noted from these studies that pot milling and ball milling proved best.

Because of its specificity, the

PRETREATMENT OF NEWSPAPER

| Newspaper (Boston Globe) | % Saccharification 1 hr | 4 hr | 24 hr | 48 hr |
|---|---|---|---|---|
| Mighty Mac - Mulcher | 10 | 24 | 31 | 42 |
| Jay Bee - Paper Shredder | 6 | 12 | 24 | 27 |
| Pot Mill | 18 | 49 | 65 | 70 |
| Sweco Mill | 16 | 32 | 48 | 56 |
| Granulator-Comminutor | 6 | 14 | 24 | 26 |
| Fitzpatrick (Hammer Mill) | 10 | 16 | 25 | 28 |
| Majac (Jet Pulverizer) | 11 | 15 | 26 | 29 |
| Gaulin (Colloid Mill) | 9 | 17 | 27 | 31 |
| Soaked in Water | 7 | 13 | 24 | 28 |
| Boiled in Water | 4 | 9 | 21 | 26 |
| Treated 2% NaOH | 8 | 14 | 28 | 35 |
| Viscose | 15 | 30 | 44 | 51 |
| Cuprammonium | 18 | 35 | 52 | 58 |

cellulase enzyme reacts solely with the cellulose and is not affected by other impurities present in the reactor. Figure 5 shows the results achieved with milled newspaper digested in a stirred tank reactor. (15). Glucose syrups of 2 to 10% concentrations were realized. The ink, lignin, and other impurities present did not cause any problems. The residue after hydrolysis was a black sticky material that dried to a hard unwettable cake. This material is chiefly lignin which can be burned as a fuel or used as a source of chemicals.

HYDROLYSIS OF MILLED NEWSPAPER IN STIRRED REACTORS

| Enzyme Protein mg/ml | Newspaper % | Temp C | Glucose 1 hr % | 4 hr % | 24 hr % | 48 hr % | Sacchar-ification % |
|---|---|---|---|---|---|---|---|
| 0.7 | 5 | 50 | 1.0 | 2.0 | 2.8 | - | 50 |
| 0.7 | 5 | 50 | 1.0 | 2.0 | 2.3 | - | 42 |
| 1.0 | 10 | 50 | 2.1 | 3.1 | 5.5 | 7.3 | 66 |
| 1.6 | 10 | 45 | 2.0 | 3.6 | 5.4 | 6.5 | 59 |
| 1.6 | 10 | 50 | 2.3 | 4.2 | 6.4 | 6.3 | 57 |
| 0.8 | 15 | 45 | 1.5 | 2.8 | 5.3 | 7.7 | 46 |
| 0.8 | 15 | 50 | 0.8 | 2.8 | 6.1 | 6.3 | 38 |
| 1.8 | 15 | 50 | 3.2 | 6.0 | 8.6 | 10.0 | 60 |

Reactor Volume 1 Liter Stirred 60 RPM pH 4.8

Having proved that this process is technically feasible, our next step is an intensive pilot plant study to optimize all variables and obtain the engineering and economic data needed for the design of a demonstration plant.

In collaboration with Fermentation Design, Inc. of Bethlehem, Pa., we have engineered a highly instrumented pilot plant consisting of such equipment as:

1. Fermenters
2. Enzyme reactors
3. Holding tanks and auxiliary vessels
4. Instrumentation modules
5. Substrate handling and preparation equipment
6. Enzyme recovery and concentration equipment

The design and construction is such that the most sophisticated fermentation techniques including batch, continuous and semi-continuous processes can be studied.

Because of the sophistication of the monitoring and control instrumentation, both the fermentation and the enzyme hydrolysis will be continuously monitored and controlled in order to optimize the output of the individual processes. Figure 6 shows the 250 liter biological reactor that will be used to study the cellulose hydrolysis. Figure 7 and 8 show the 400 liter fermenter with its

30 liter seed fermenter that will be used to produce the cellulase enzyme from the **Trichoderma viride** fermentation. Figure 9 shows the instrumentation cabinets for

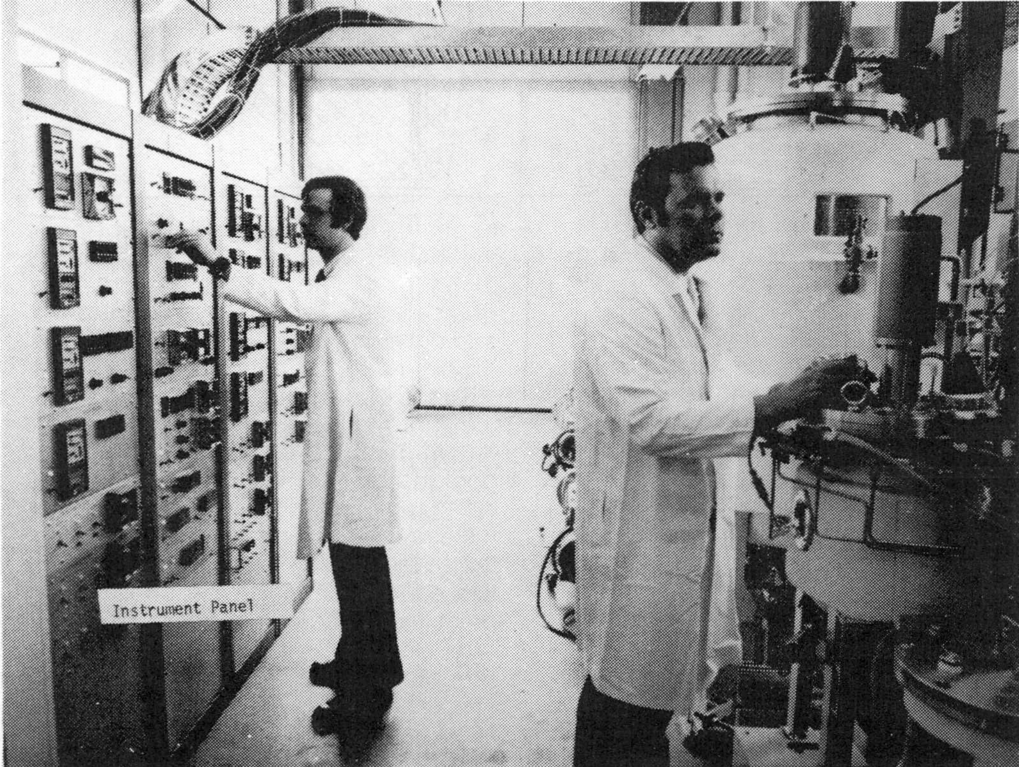


the fermenter and enzyme reactor which contain modules for control or analysis of temperature, pressure, agitation speed, pH, sparging, dissolved oxygen, vessel weight, liquid level, and exit gas.

Figure 10 shows the simplified schematic of the process. The initial capacity of this pilot plant is the processing of 1000 lbs. of cellulose per month. This equipment is now being installed at Natick and will be operational by June. Our projected demonstration plant is to handle 200,000 lbs. per month.

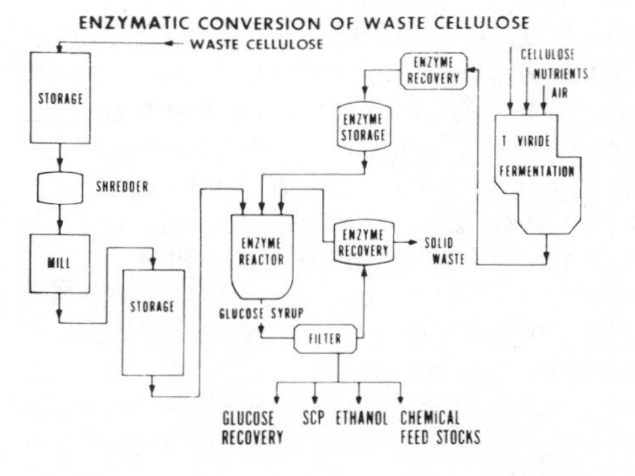


Because of the significant potential contribution this process can make to Project Independence, it has been brought to the attention of the National Science Foundation, the Atomic Energy Commission, and the Federal Energy Office.

In conclusion we at Natick are convinced that:

1. The vast quantity of cellulose that is replenished annually should be exploited as a source of energy, food, and chemical feedstock.
2. The enzymatic hydrolysis of such energy rich material as cellulose to glucose is technically feasible and practically achievable on a very large scale by 1980.
3. The use of ethanol-gasoline fuel blends to power automobiles and other internal combustion engines should be encouraged in order to extend our petroleum reserves.
4. The exploitation of our fossil fuel reserves be it coal, oil shale or otherwise, may satisfy our energy demands for the next five to ten decades. However, what energy source can we explore at that time?
5. We believe that the ultimate long range solution to the world's energy problem is the development of practical and economical processes capable of harnessing the inexhaustible energy of the sun.

We at Natick Laboratories look forward with great expectation and confidence to the opportunity of contributing to the effort that will make Project Independence a reality.

### References

1. H. C. Hottel and J. B. Howard "New Energy Technology" (MIT Press, Cambridge, 1971) p4.
2. J.A. Bolt, "A Survey of Alcohol as a Motor Fuel" SAE Conference Paper SP254, June 1964 (65 references)
3. R. K. Pefley, M. A. Sand, M. A. Sweeney, and J. D. Kilgore, "Performance and Emission Characteristics using Blends of Methanol and Dissociated Methanol as an Automotive Fuel", Paper 719008, 1971, IECEC, p36.
4. Reese, E. T.; Mandels, M., and Weiss A. N. 1971. Cellulose as a novel energy source in Advances in Bioengineering. 2. Ed. T. K. Ghose, A. Fiechter, and N. Blackbrough. Springer Verlag. 181-200.
5. Katz, M.; and Reese, E. T. 1968. Production of Glucose by Enzymatic Hydrolysis of Cellulose. Applied Microbiol. 16: 419-420.
6. Mandels, M.; and Weber, J. 1969. The Production of Cellulases. Adv. Chem. Series 95. 391-414.
7. Ghose, T. K. 1969. Continuous Enzymatic Saccharification of Cellulose with Culture Filtrates of **Trichoderma viride** QM6a. Biotech. Bioeng. XI 239-261.

8. Ghose, T. K.; and Kostick, J. 1969. Enzymatic Saccharification of Cellulose in Semi and Continuously Agitated Systems. Advances in Chem. Series 95. 415-446.
9. Ghose, T. K.; and Kostick, J. 1970. A model for continuous enzymatic Saccharification of Cellulose with Simultaneous Removal of Glucose Syrup. Biotech. Bioeng. XII 921-946.
10. Mandels, M.; Weber, J.; and Parizek, R. 1971. Enhanced Cellulase Production by a mutant of **Trichoderma viride**. Applied Microbiol. 21: 152-154.
11. Mandels, M.; Kostick, J.; and Parizek, R. 1971. The use of absorbed cellulase in the continuous conversion of cellulose to glucose. J. Polymer Sci Part C, No. 36: 445-459.
12. Ghose, T. K. 1972. Enzymatic Saccharification of Cellulose. U.S. Patent 3,642,580. Feb 15, 1972.
13. Mandels, M.; Hontz, L. and Brandt. 1972. Disposal of Cellulosic Waste Materials by Enzymatic Hydrolysis. Army Science Conference Proceedings June 1972. Vol. 3. AD 750 351: 16-31.
14. Mandels, M.; and Kostick, J. 1973. Enzymatic Hydrolysis of Cellulose to Soluble Sugars. U.S. Patent 3,764,475. Oct 9, 1973.
15. Brandt, D.; Hontz, L.; and Mandels, M. 1973. Engineering Aspect of the Enzymatic Conversion of Waste Cellulose to Glucose. AIChE Symposium Series 69. No. 133. p127-133.
16. Mandels, M.; Hontz, L.; and Nystrom, J. 1974. Enzymatic Hydrolysis of Waste Cellulose. In preparation.

# A Mini Methane Generator

One of the nice things about methane generators is that they can be built in almost any size. Here is a description of one that you can build yourself from easily available materials. It won't generate very much gas, but it will show how the process works, and demonstrate that biologically produced methane really will burn.

The digestion takes place in a gallon glass jug. The jug sits in a bucket full of water with an aquarium heater/thermostat to keep it at the right temperature for best gas production. The gas is stored by water displacement in another glass jug. When a pinchcock is opened, water forces the gas through a jet where it can be ignited.

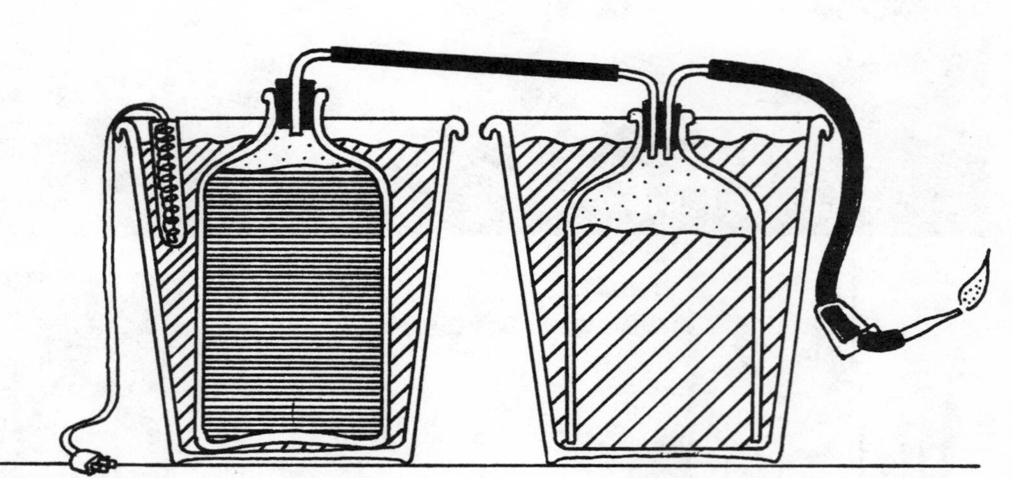

You will need two buckets, two gallon jugs, an aquarium heater, a thermometer, two rubber stoppers (one one-hole, one two-hole), a pinchcock, some rubber tubing, and some glass tubing. The last four items should be available in any large hobby shop.

While you are getting the parts of your apparatus together, you can prepare some starter brew. It will save a lot of time in getting your experiment going. Get a bottle and put in a fresh sample of every sort of dung you can find, some septic tank sludge, some pond bottom muck, etc. Fill with water to within a couple inches of the top. The idea is to get as many different kinds of anaerobic bacteria together as possible and give them a chance to get used to conditions inside a digester. The ones that like it in there will multiply rapidly and be all set to get things off to a quick start when you fill your digester. Do not seal the bottle tightly. Cap it loosely enough so that gas can escape and give it a shake every day or so.

You will need to cut the bottom out of the jug you are going to use for a gas collector. Any one of the bottle cutters around will do a neat job. If you can't get ahold of one, score the jug evenly all around with a glass cutter. Then heat the scored line all around with a candle flame. Now take an ice cube and run a corner of it along the score. The glass should crack right along the score as it comes into contact with the ice cube. When the crack is all the way around, the bottom should come right off with a little encouragement. A ragged edge is perfectly acceptable. If you botch the job, just get another jug and try again.

Now you have to cut and bend some of that glass tubing. To cut the tubing you need a small three cornered file. With the file make a small scratch in the tubing where you want the cut to be. Turn the tubing so that the scratch faces away from you, place your thumbnail opposite the scratch, and bend outwards. The glass should snap cleanly at the scratch.

For safety the freshly cut ends of glass tubing should be fire polished. This is done by holding the end in a gas flame (your kitchen range with the flame turned high will do fine) long enough for the heat to melt and round the sharp edges. Turn the tubing slowly so that it will heat evenly. Don't hold it there too long, or the tubing will melt itself shut. You will need two short pieces of tubing, each about three inches long, to connect the rubber tubing and the stoppers. Making them first will give you some practice in cutting and fire polishing tubing.

Next you need to make a 90° bend in a piece of tubing. Take a piece at least eight inches long and, holding it by its ends, rotate it slowly in the flame. Keep it moving so that at least three inches of the middle is heated. After a while, it should begin to feel rubbery and start to sag in the middle. When this happens, help it along. Don't force it. When the bend is great enough, take it away from the flame and put it down to cool. Then cut off any excess tubing and fire polish the ends.

The last part you will need to make is the flame jet for burning the gas. Heat a length of tubing as though you were going to bend it, but when the glass starts to soften, pull gently on the ends instead. The soft part will stretch like taffy. When you have pulled the tubing in two, each piece will have a long tapering point like a medicine dropper. (And how did you think they made medicine droppers?) When they have cooled, pick one, cut the very tip off, and hold it briefly in the flame to fire polish it. Be careful not to seal the end shut. That completes the glassworking part of the project. If any of it gives you trouble, get the help of a chemistry student.

Now you can get ready to fill the digestor jug. Mix together your starter, some manure, and some water. Stir the mixture until it is free of lumps and add water until it has the consistency of thick cream. Fill the jug about three quarters full. A plastic bottle with the bottom cut off makes a good disposable funnel. Assemble the rest of the apparatus. Shove the gas collector jug under water to force the air out of it before connecting the stoppers. Set the thermostat and check the thermometer from time to time to make sure the temperature is in the 80-90° F range.

You can see that gas is being generated when the water level in the storage jug begins to drop. If enough gas is produced to force all of the water out of the jug, the excess will bubble out and escape into the air. This automatically keeps the pressure from becoming excessive. The first three days or so, the system will produce large amounts of gas. This will be mostly carbon dioxide. If you try to light it, it will probably blow out the match instead. Methane production will start in two or three weeks, depending on how long it takes the bacteria to settle down and get to work. The first time the gas actually ignites and burns is a real thrill!

Keep day by day records of the gas produced and the water temperature. You will find them to be of great value later on. If at any time you suspect that some air has gotten into the system, vent all the gas and start over again. The amount of gas involved is small, but there is no point in taking chances.

When production stops, save some of the spent solids to use as starter for your next batch. When you have established a good working mix of bacteria, you can start experimenting with different temperature settings and new mixtures of manures and vegetable materials.

# A Homesite Power Unit

By Les Auerbach

The following is an account of a methane generator built by Les Auerbach and Ben Katz. It is presently functioning as a demonstration unit at the University of California.

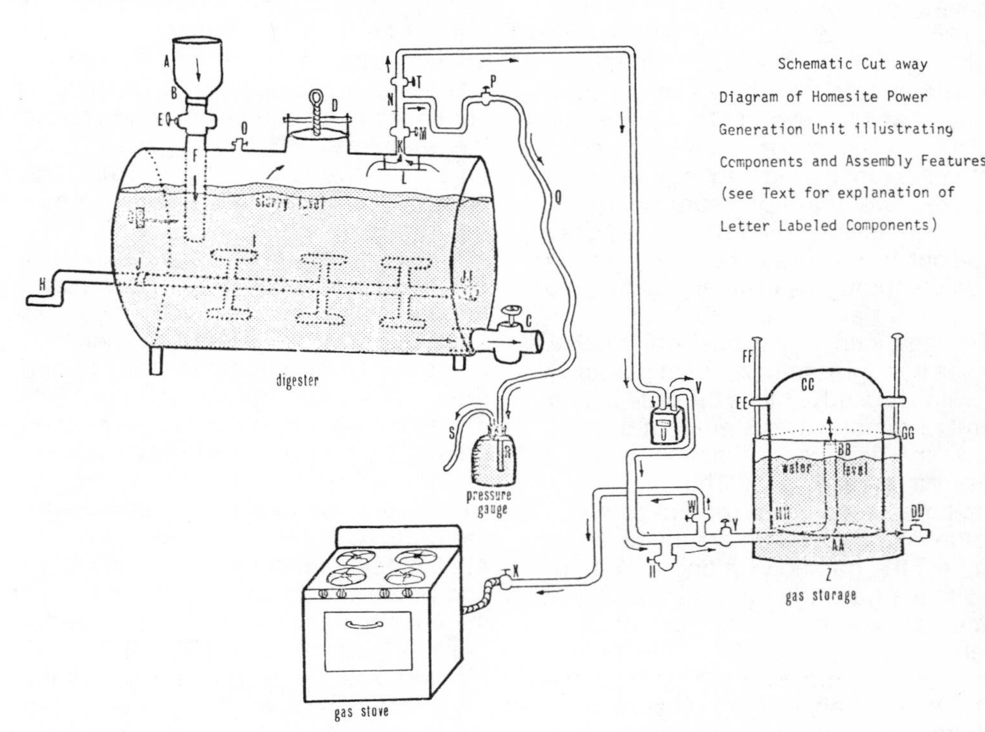


The digester is a steel tank 66 inches long, having a diameter of 45 inches, located one foot above the ground on about a 5 degree incline in the direction of the 4 inch slurry exit valve (C). Since the unit was located in a moderate climate, the above ground digester used the heat from the sun for creating optimal temperatures. The incline toward the exit valve allows for easy slurry flow out of the digester. The digester is equipped with a hatch (D), 18 inches in diameter directly on top and at a point midway between either end of the tank. The loading apparatus is made of a drum (A), 22 inches in diameter and 34 inches high, to which a funnel (B), 22 inches in diameter at one end and 4 inches in diameter at the other, is attached. The loader is secured by way of a 4 inch galvanized union to a 4 inch gate valve (E) used as the continuous feed inlet to the digester. Extending 6 inches above the tank and attached to the valve (E) is a 4 inch diameter pipe (F) extending 2 feet into the tank. This pipe accepts new organic matter, empties directly into the existing slurry, and prevents the gas located in the upper part of the tank from escaping during loading operations.

A thermometer (G), capable of measuring temperatures from 40 degrees F to 120 degrees F is located toward the middle of the tank about 18 inches from the bottom. The thermometer is equipped with a sensor unit that extends 12 inches into the slurry in the digester. This gives an accurate temperature reading of the slurry, providing a cover is placed over the thermometer when exposed to direct sun. The digester is equipped with a hand operated stirring mechanism (H). This stirrer consists of a 1 inch diameter steel rod inside the digester and three paddles attached at intervals of 15 inches. There are two blades (I) on each paddle extending out and away from the steel rod a distance in length equal to a point about 2 inches from the bottom of the tank. A crank extending out from the end of the tank 18 inches, is attached to the steel rod. A compression fitting (J) attached to the end of the tank through which the rod passes, provides a tight seal.

A gas valve (O) is located on top of the tank for purposes of testing the system for leaks, with air under pressure. A baffle consisting of a flat piece of metal 8 inches wide and 6 inches long (L), is suspended by two brackets (3 inches in length, 1½ inches wide) from the inside of the top of the tank directly under the gas outlet pipe (K), and is designed so that the gas must flow up around and then out the exit pipe. This baffle prevents any particles from being carried upwards with the evolving gas, clogging the pipes and causing other unwarranted complications.

The gas passes through the ¾ inch diameter exit pipe (K) past a valve (M) which is used to shut the gas off from the digester. The gas continues to a point in the line (N) where a ¾ inch diameter tee reduced to ⅜ inch on one end, is placed. The ⅜ inch diameter pipe comes off the tee for 6 inches, at a 90 degree angle, then down 12 inches, horizontal again for a distance of 4 inches, and back up 12 inches where a ⅜ inch gas pet cock (P) is located. This small line with its jog can be used for a pressure release valve. Filling it with water and leaving the pet cock(P) open creates a column of water 12 inches high. A 12 inch column of water is equal to .432 lbs./sq. in. (note 3) of pressure, which means that if the system, for any unforeseen reason, should climb above .432 lbs./sq. in. of pressure, it would blow the water out of the ⅜ inch line, allowing the gas to escape, leaving no immediate danger to nearby areas.

Although this is a common way of building a pressure release valve, a refinement, if so desired, that is equally effective and easy to install, can be attached at point (P). Do not fill the ⅜ inch line with water. Attach a flexible ⅜ inch hose (Q) at point (P); extend it to a one gallon jar of water through one hole of a double hole cork. Attach a brass or other piece of non-corrosive ⅜ inch pipe, 1 foot long, to the end of this flexible hose and into the water filled jar. Through the other hole in the cork (but, not down into the water) pass a 4 inch long 3/8 inch brass rod with another piece of flexible 3/8 inch hose. Mark lines, with intervals of 1 inch, on the outside of the jar. This jar is a pressure gauge measuring in inches of water with the capability of releasing excessive amounts of pressure. By moving the hose with its rigid end (R) up and down in the 11 inch high water filled jar, a measure of gas pressure in the system in inches of water can be obtained.

The pressure in the system is equal to the water level because when gas bubbles come out of the end of the rigid pipe (R), the incoming gas will have to be slightly greater in pressure than the pressure created by the water depth at the level where the end of the brass rod is. This simple gauge is an efficient system for monitoring and regulating the gas pressure. If the pressure builds above the desired level so that gas does come out the end of the rod, it will bubble up safely through the water and out into the open air along hose (S).

The gas, continuing past point (N), passes through another 3/4 inch gas valve (T) which is used to close the digester off from the remaining system, but still allows the digester to function with a pressure release valve. This is sometimes used when repairs or additions to the system are made. From point (T) the gas flows along a 3/4 inch pipe into a gas meter (U), similar to that placed outside every establishment where natural gas is used. This meter can measure from 1/2 to 1,000,000 cubic feet. The gas, after being measured, exits the meter (V) and continues to a place (W) in the line which branches, going to the gas storage unit or directly to some gas use. In this case, the gas moves to a gas stove, passes through a valve (X) before entering the 1/2 inch flexible connector delivering it for use.

The water trap (II), placed directly before the branch (W) (which is nothing more than a piece of 2 inch pipe directed downward with a valve at the end), can be opened periodically to drain the collected liquid.

The gas at the branch off (W) would also continue past the gas valve (Y) used for shutting off gas supplies to the gas holder. The pipe at a point (Y) 12 inches from the gas storage unit, becomes 1 1/2 inches in diameter and enters the gas storage unit 2 inches from the bottom of the tank (Z) holding the water, continuing in 2 feet to the center of the tank making a 90 degree angle upward (AA). The pipe continues to a height of 3 inches above the water level (BB) where gas exits and enters the inverted drum (CC) 4 feet high, 45 1/2 inches in diameter and in which the gas is stored.

In case, for any reasons, water must be emptied from the tank (Z), it can pass through a 1 inch valve (DD). The outer tank (Z) has a diameter of 51 inches, being 5 1/2 inches larger than the inverted drum (CC) allowing for a space of 2 3/4 inches all the way around. The inverted drum (CC) rides up and down depending on whether gas is entering or exiting the storage unit. Guides (EE) are attached to the drum (CC) through which rods (FF) 4 1/2 feet tall, 3/4 inches in diameter, pass and which are secured at the top (GG) of the outer water drum (Z). These guides prevent the two drums (CC, Z) from binding. A situation could arise where the gas was produced at a greater rate than being used. In such a case, the drum (CC) would be fully extended and the four 1/2 inch diameter holes, (HH), placed at a distance above the bottom of the drum (CC) so that when the drum is fully extended they will lie above the water level, and allow the gas to exit into the atmosphere. It should be mentioned that all gas pipelines have incline of 1/4 inch to the foot, either toward a receptacle containing water, slurry or water trap. This allows for any condensing moisture in the gas (from either the digester or storage tank) to flow down the pipe to a draining area and not block the flow of gas.

The gas flowing toward the storage unit can either go directly to a use or continue on into the storage unit. The gas coming back from the storage area to the branch (W) can also continue on to a consumption device.

A gas scrubber, which contains ferric oxide filters for excluding $H_2S$, or lime water solution for limiting $CO_2$ concentrations, can be installed between the digester and the gas meter. All that is necessary is to run the main gas line from the digester into the bottom of a sealed drum containing the oxide filters (or a solution of lime water) and let the gas rise up to the top where it exits through a pipe continuing on through the gas meter to the storage facility. The lime water can be emptied and refilled when needed and the ferric oxide filters exposed to the air can then be replaced after use. The lime water solution can be used in the soil without contaminating it. A gas scrubber to remove $H_2S$ is only necessary if the gas is used in a combustion engine where it could pit metal cylinders.

Complete details of the design and construction of this methane generator can be found in "A Homesite Power Unit: Methane Generator" available for $5 from Les Auerbach, 242 Copse Road, Madison, Conn. 06443.

# Backyard Bio-Gas Plant

**By Sharon Whitehurst with James Whitehurst Ralph Hurd**

If you're into energy and ecology, if you're homesteading, gardening, or just trying to beat the high costs of conventional living, you need a bio-gas plant!

We built our own plant for converting organic wastes to methane gas and beautiful fertilizer in the summer of 1972 under the expert guidence of India's Ram Bux Singh. We're so excited about the bio-gas concept and its many possibilities, that we'd like to share our experiences with you. We're going to assume for the purposes of this article that you've already done some background reading on the subject and have an understanding of the fundamental bio-gas process.

There are many structural modifications possible in the building of a bio-gas plant, but all plants whether large and sophisticated or small and simple, have in common the three following components: a container in which the actual digestion of raw materials takes place, a gas dome or gas collector (in large models sometimes this is a separate structure) and an agitator.

Various organic substances or combinations can be digested efficiently in a bio-gas plant as long as a carbon nitrogen ratio of 25-1 is maintained.

The bio-gas plant which we built here at Ral-Jim Farm is a simple 225 cubic foot model chosen by Ram Bux Singh to demonstrate the process. He felt this type of plant could be most easily assempled and operated. It is the size commonly used by India's typical farms with a "herd" of three or four cows, and is a nice size for any small farm of homestead set up.

Although we have used only cow manure in our experiments, this same plant can be fed with other wastes.

The first concern after we decided to built a bio-gas plant was fairly obvious: where to put the thing! If the main use of the methane gas is to be household cooking the simplest place to install it is outside the kitchen. That way you can pipe the gas into the house, directly from the plant's gas dome. If you're lucky enough to have a kitchen with a southerly exposure, so much the better, as the sun's warmth will facilitate gas production during winter. Our plant had to be stationed on the north side of the house, and in our severe Vermont winter's gas production drops substantially—a crude shelter and auxilary heat would be necessary to maintain even year round gas production.

The single most costly component

($200.00) in our plant was an old iron boiler 5½ ft. in diameter by 16 ft. deep which will hereafter be referred to as the TANK. (You can use any sturdy, non leaking container of this approximate size. The main thing to remember is to figure things so that the diameter of the gas dome is six inches less than the diameter of the tank.)

Two angle iron **cross braces** must be welded across the inside of the tank, one at a depth of 4½ ft. from the top, one at six feet. These are positioned slightly off center as their purpose is to hold a **center pole** in place. The center pole is 2½ inch pipe centered in the tank welded to the cross braces. It must extend 4 ft. above the top of the tank. (See diagram A.) The second step in the construction of our digestor was the **heating coil** used to circulate hot water (recycled from bath tub or automatic washer) around inside the tank, thereby heating the raw materials for faster digestion. If you live where it's warm year round, you can skip the next paragraph!

The type of heating coil which we used was rather a pain to put together and anyone who can think of something easier has our blessing! We bought about 55 ft. of one inch galvanized pipe which we took to a local plumbing establishment where we had the pipe bent into a 4½ ft. spiral form. (This proved to be a rather costly manuever; we suggest you investigate the various types of flexible plastic pipe on the market, something which could more easily be bent to shape.) We fastened our finished coil to a series of braces, six inch pieces of angle iron which had been welded at 6 ft. intervals around the lower part of the tank in a circular pattern. (Diagram B) An inlet and outlet were made by installing six inch fittings with elbows through the side of the tank at the appropriate sites to connect with the top and bottom of the heating coil. The hot water enters the coil at the bottom, and exits at ground level from the top of the coil. When the plant is all assembled on your chosen site, make a drainage ditch, lined with stones or proper drainage tile to carry the water away from the immediate area of the digestor. How to transport the waste hot water from house to digestor is a problem which will be fairly unique to each builder, and will involve a certain amount of under ground plumbing for cold weather use. If you've stuck your plant out behind the stable or at the far end of the garden, then you've really got problems!

If you plan to use your digestor on a continuous feed basis, you will at this point prepare for charge and discharge fixtures, mixing and holding tanks. We had originally planned to use our plant as a continuous feed unit, but later decided that batch feeding was more practical (less work!) for our purpose of demonstrating the plant to the public.

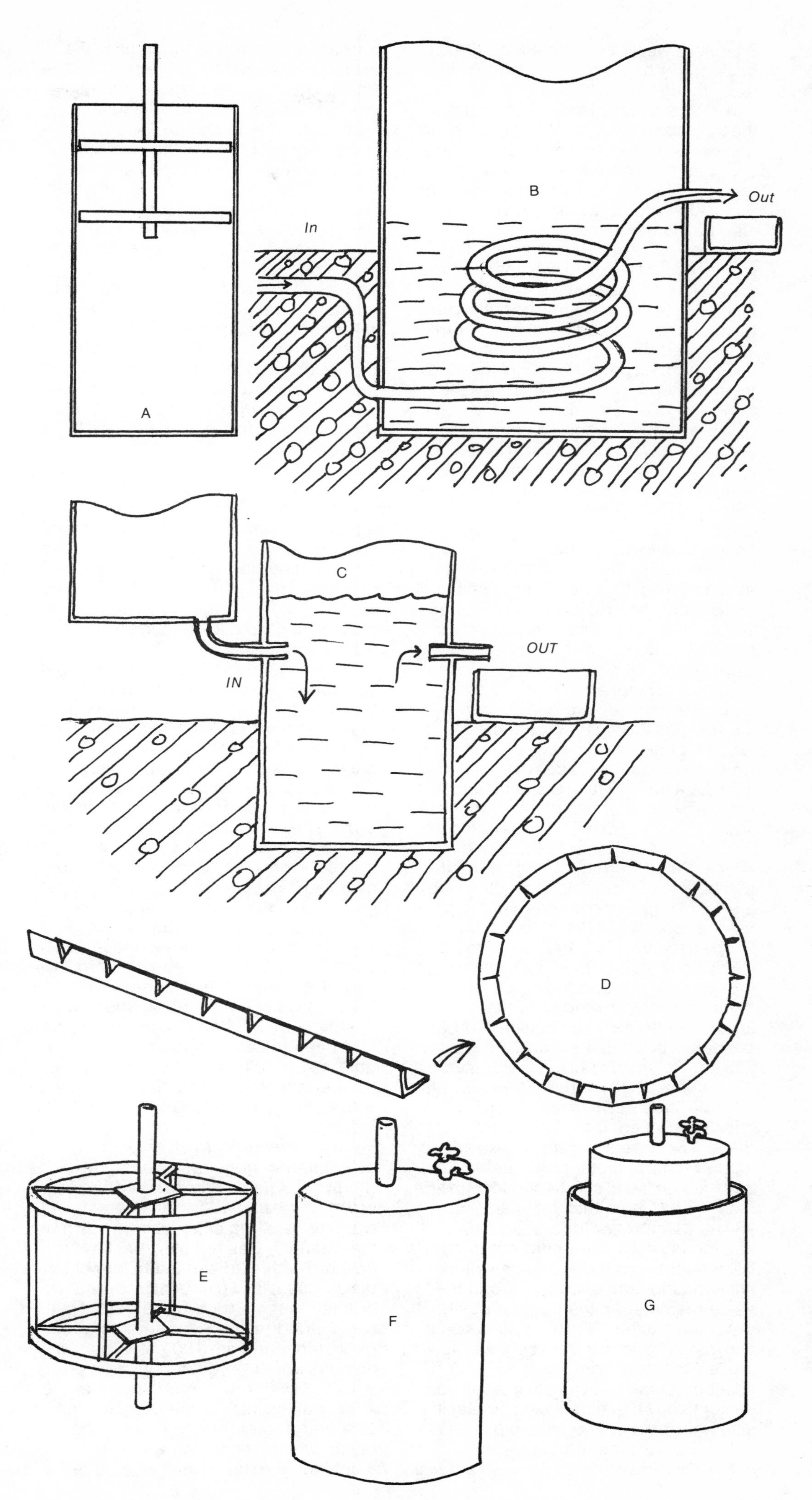

Therefore we didn't ever set up the mixing and holding tanks. The general idea, though is as follows!

We cut a 4 inch circle in both sides of the tank about six feet down from the top and welded a piece of metal pipe in at a slight angle. This formed a "collar" to receive a length of 4 inch rigid plastic pipe. After the digestor was set up on site, two large containers would be arranged—one in which fresh organic matter would be mixed with the required warm water, set **above** the inlet pipe, the other, would be set **below** the outlet pipe to catch the digested material or **slurry**. When used on a continuous feed basis the system will be self leveling, that is, as fresh matter which is heavier is sent down the inlet to the bottom of the digestor, lighter digested slurry will automatically be forced up to run out into the collecting tank. (Diagram C.) We suggest you see Ram Bux Singh's booket "Bio-Gas Plant Generating Methane from Organic Wastes" for more specifics on continuous feed plants.

Constructing the gas dome was the next and most time consuming phase of our project, and incidentally, required some rather touchy welding on very thin metal. The **dome frame** was the first step. Angle iron was cut into two five foot lengths and notched at intervals so that it could be bent and welded to form two circles. (Diagram D) Three more four ft. lengths of angle iron were then cut and welded vertically between the two hoops to make a frame 5 ft. in diameter and 4 ft. high. Next we cut two **plates** 8 inches square from ⅛ inch thick metal and cut a 3 inch circle out of the center of each. These plates were attached, to the top and bottom of the frame respectively, by means of angle iron braces welded from the four corners of the plates. (Diagram E) The assembly is to hold a **guide pole** which consists of 4 ft. of 3 inch metal pipe which is run through the holes in the metal plates and welded to them. This allows the dome when finished to be slipped over the center pole of the tank to rise and fall on the pole, as regulated by the volume of gas within.

Three sheets of mild steel 4′ x 8′ x 1/16″ thick were needed to complete the gas dome. Two sheets of steel were wrapped around the perimeter of the frame (the extra foot of metal positioned to form a 6 inch overlap on either side) and the seams carefully welded. Extreme caution must be used here not to burn holes in the thin metal, which would allow gas to leak out later. A cap the same diameter as the frame plus about two inches extra all around for a "seam allowance" and a 3 inch center hole was pieced together from the remaining sheet of metal. (Obviously it would be easier to buy the third sheet of metal in a size which would allow the cap to be cut out whole, eliminating the seam.) The seam allowance is bent down all around and carefully welded into place. We added a small metal handle on top of the dome. By grasping it, the whole dome can be rotated in the tank, causing the vertical braces to agitate the decomposing matter in the upper part of the tank. A small faucet fitting was set into the dome cap several inches from the edge, so that the gas supply could be "tapped". (Diagram F finished dome) The seams of the dome were daubed with liquid steel and the dome was painted inside and out with several coats of enamel paint.

Frequent stirring or agitation of the stewing matter in the tank is needed to break up the mat of scum which forms on the surface and slows down digestion. We made an **agitator** of an old brake drum attached to a length of ¼ inch thick metal cable. When assembling the plant the cable is threaded up through the center pole of the tank with several feet of cable left over. We manipulated this agitator by standing on the dome and hauling on the cable to churn the brake drum up and down through the slurry. A better way might be to allow extra cable which could be rigged with a pulley so that one could stand on solid ground and work the agitator.

If you've followed through to this point you should have something resembling the now famous Ral-Jim Farm bio-gas plant (Diagram G) ready to assemble on your chosen site. We had planned to set our unit underground to within 4 ft. of the rim of the tank, but the water table was discovered to be too high to permit this on our site, so seven feet of the tank (plus the gas dome) are above ground. We poured a three inch thick pad of cement over the slightly squelchy bottom of our excavation, let it dry and laid down a sheet of black plastic, then a floor of boards, covered with a second sheet of plastic. The tank was then lowered gently into place and another three inches of cement poured in to seal the whole works solidly in place. When the agitator cable had been threaded through the center pole and the dome settled over it, the extra space in the excavation was filled with old hay, leaves, saw dust etc. for insulation.

About a week before the plant was scheduled to go into operation, Mr. Singh half filled a fifty gallon drum with cow manure and added an equal part of warm water. This concoction, known as the "starter", was agitated briskly, covered with a lid and left to ferment in the July sun. It also had to be stirred two or three times a day. The digestor tank was loaded (using a farm tractor and bucket loader) with 100 cu. ft. of cow manure which we diluted with an equal volume of warm water. (10 tons manure, 1,000 gal. water) For this job we simply attached a long garden hose to the faucet which serves the washing machine. The starter mix was added to the tank and agitated several times a day. Within four days the gas dome began to rise, almost imperceptably at first, then very noticeably as gas production got under way. The day when we first lit a flame in the gas stove burner connected to the dome faucet with flexible hose was a high day indeed!

Last summer was the first chance we had to witness first hand what bio-gas slurry can do for a plot of earth when used as fertilizer. For years we had hacked away at our hard clay soil, meanwhile envying those gardeners blessed with more workable dirt. Singh predicted miracles, and as far as we're concerned they happened. Since our garden spot is only a few feet behind the digestor, we hoed a series of little irrigation ditches and pulled the plug on the outlet pipe. A river of digested slurry, the consistency of homemade pea soup flowed onto the garden. We harrowed it in and proceeded to raise a real show crop of vegetables. Our unwieldy soil softened and loosened, becoming friable and easy to work. The few weeds pulled out whole instead of snapping and landing the incautious weeder on his rear as had been the norm in other years.

In the two years since we set up our plant we have done the "show and tell" thing for hundreds of visitors, news reporters, and environmental groups. We have cooked on our methane burner, and run a John Deere tractor on methane. Thousands have written us requesting Ram Bux Singh's $5.00 book on methane making. We have plans and blue prints for the large bio-gas plant which could utilize the waste from a herd of one hundred animals to make a farm of this size virtually independent of conventional (and costly) energy. We had hoped that financial aid for this big project could be had from some governmental or private research agency, but so far a lot of bright leads have fizzled out, for us anyway.

However many folks at the grass roots level are experimenting, reading and starting to build their own bio-gas plants. This is heartening. A concept with as much practical potential should not lie idle. More successful backyard plants such as ours will help to propagate the idea and prove its effectiveness to the general public.

This article has only touched on our experiences, so much more could be said about the chemical process, the various types and plants possible, etc. You have probably gathered that one can be somewhat innovative in the materials used for a small digestor. We found the local junk yard to be a great source. This is a project which most any handy man can undertake successfully. We hope many of you will be inspired to do just that!

# Geothermal Energy

Geothermal energy is the use of heat from the depths of the earth to generate power. As early as the 17th century it was noticed that the deeper a mine shaft was dug, the warmer it became. Actual use of geothermal energy is as old as the practice of bathing in hot springs. In 1904 natural volcanic steam was used to spin turbines and generate power at Larderello, Italy. The idea of harnessing geothermal heat as a practical power source, however, is just beginning to attract widespread interest.

The only geothermal power plant in the U.S. today is The Geysers, north of San Francisco. This area first called attention to itself through its natural steam vents. The first attempt to tap its steam was made in the early 1920's. The drilling was successful, but the local demand for electricity was too small to justify the building of a power plant. Starting in 1955 more holes were drilled and the first power plant was built in 1960. Steam directly from the wells drove turbines which turned generators to produce electricity. By 1970, 75 wells had been drilled, of which 70 produced steam. Today 100 wells supply enough steam to generate 411 megawatts.

The steam from the wells averages 100 pounds/square inch (psi) at about 205°C. This compares with steam of 3000 psi at 550°C used by modern fossil fueled power plants. Because of the lower temperatures and pressures, the turbines at the Geysers are roughly one third less efficient than the turbines in a fossil fuel plant. It takes 450 MW of geothermal heat to produce 100 MW of electricity. The waste heat is disposed of by evaporating most of the condensed steam in evaporative cooling towers. The 20% remaining is pumped back into the ground to recharge the source and to prevent the mineral-bearing water from contaminating local streams.

New wells are being drilled all the time and present plans are to increase generating capacity by about 100 MW per year. The Geysers may be producing 1,000 MW by 1980. The ultimate potential of the field may be as high as 3,000 MW. The limits of the field have not yet been determined. It is thought to cover at least twenty square miles and may well be the largest in the world.

The source of geothermal heat seems to be radioactivity. Nearly all rocks contain faint traces of radioactive elements. As these break down, a small amount of heat is released. At the surface this heat is quickly lost to the air, but deep in the earth's interior it seeps away so slowly that it has built up to rock melting temperatures. In effect, geothermal heat is atomic heat from a reactor as big as the earth.

The amount of energy contained as heat in the earth's interior is enormous. It has been calculated that cooling the earth by one tenth of one degree Celsius would supply the world's energy needs at the present rate for four million years!

Unfortunately it is possible to tap only a small part of this energy at the few places that an upthrust of magma has left hot rock close to the earth's surface. While it is theoretically possible to drill as deep as four or five miles in search of geothermal heat, it is obvious that the less drilling, the more economical the power. A geothermal resource is therefore an area where high temperatures are found relatively close to the earth's surface. Geothermal fields occur chiefly near regions of high earthquake activity. At these points the heat flow from the inside of the earth is up to ten times the average.

There are four main types of geothermal field. The dry rock field is simply an area which has dry hot rocks close to the surface. None of these sources has been commercially developed. It has been theorized that cold water could be injected into the hot strata to crack the rock. The water would then be heated to the boiling point and steam could be drawn from a second bore and used to generate power. Work on this idea is presently going on at Jemez Plateau near Los Alamos, New Mexico, where test holes have been bored into the granite base of an ancient volcano. It has already been demonstrated that the hot rock can be fractured by pumping in cold water at pressures of 10,000 to 15,000 psi. Leakage is small. It is not known yet if the technique will be an economical means of producing power. The costs of drilling must be spread over the bore's active life, which is not known, thought it is suspected to be twenty years or more.

The dry steam type of geothermal source is the best known because it is the easiest to exploit. It consists of a layer of hot, porous rock capped with a dense layer of impermeable rock. Water seeps into the porous rock and is turned to steam by the heat. The upper layer of rock prevents the steam from escaping, forming a natural boiler. When a hole is bored through the top layer of rock, a tall plume of superheated steam goes roaring into the air under high pressure. The fields at Larderello, Italy; The Geysers, California; and Valle Caldera, New Mexico are of this type.

The wet steam type of geothermal field is about twenty times more common than the dry steam type. It contains a large amount of water and relatively little steam under high pressure. When the pressure is relieved by drilling, part of the water flashes into steam, forcing a mixture of steam and hot water to the surface. The discharge is 10-20% steam, the rest water. The geothermal fields at Wairakei, New Zealand; Parantunka, U.S.S.R.; Cerro Prieto, Mexico; and the Imperial Valley, California are of this type.

The hot water geothermal field is one in which the temperature is not high enough to produce steam. While not directly useful for producing power, the hot water from these sources can be used for home heating and many other purposes. Hungary and Iceland both have extensive fields which they have used for this purpose.

Putting geothermal energy to work is a task that requires no great technical advances or scientific breakthroughs. Geothermal wells are drilled with standard oil drilling equipment and methods. The turbines and generators that extract energy from geothermal steam are little different from those found in fossil fuel plants. Steam from dry steam field is the easiest to use. After passing through traps to remove solid particles and water droplets, it can go directly to the turbines. Steam from wet steam fields is more difficult to handle. It contains dissolved minerals that can corrode or clog pipes and tubing, and which wreak havoc with finely machined turbine blades. One solution is to pass the steam through a heat exchanger where it gives up its heat to a secondary fluid such as freon or butane which then drives the turbines. Another advantage of this method is that it can be used to produce electricity from hot water fields. There is a plant of this type in Kamchatka, U.S.S.R., which uses water at only 81°C to generate power.

Uses of geothermal energy are not confined to the generation of electricity. Hot water from geothermal wells has been used in Iceland since 1928 to heat homes. Today a majority of the homes in Iceland have central hot water heat provided in this manner. Homes in Boise, Idaho and Klamath Falls, Oregon also use geothermal heat. Another use for geothermal heat is to increase agricultural productivity in colder regions by warming the soil and preventing frost. Geothermal steam can be used directly by industry. In Kawerau, New Zealand, geothermal steam is used in a paper mill. In Japan, geothermal steam evaporates sea water to produce table salt. Geothermal brine may one day be an important source of minerals. For many years, borax was extracted from geothermal waters at Larderello, Italy. It has been suggested that geothermal brines in the U.S. could yield lithium and precious metals. Not only the minerals, but the water itself is of value. The heat in geothermal brine can be used to desalinate it, resulting in water that can be used for irrigation. It has been estimated that the Imperial Valley in California could yield a billion acre feet of water, and power as well. Presently the Department of the Interior's Bureau of Reclamation is building a pilot plant at East Mesa. If it is successful, desalinated geothermal water will be used to replenish the waters of the Colorado River.

Geothermal energy is not without its disadvantages. First, before a geothermal source can be developed, it has to be located. Unless the source calls attention to itself through hot springs or geysers, it must be located by drilling until a strike is made. Exploratory drilling is a significant part of the cost of developing geothermal power. In this respect it resembles drilling for oil. It also resembles oil in that it is a depletable resource. Eventually the underground water is depleted, or the temperature falls. Then a new hole must be drilled. To be economically successful, a field must be able to supply steam for at least 30 years, the expected lifetime of the power plant. In another respect geothermal power resembles hydro-electric power. It is tied to a fixed site, and the site is not always conveniently located with respect to the power user. A fossil fuel plant on the other hand, can be located anywhere oil or coal can be shipped.

A geothermal power plant must operate at much lower temperatures than a fossil fuel plant. In consequence it is less efficient and generates more waste heat per megawatt of electricity produced. Getting rid of this heat may be a problem in the desert and semi-desert areas where many geothermal sources are located. Cooling water from rivers and lakes is seldom available, and the high local air temperatures make air cooling difficult. One possible solution is to evaporate part of the brine brought up with the steam.

Some geothermal fields contain small quantities of hydrogen sulfide gas ($H_2S$). This gas, with its rotten egg smell, should be familiar to anyone who has ever visited a volcanic hot spring. While a single well does not produce much of it, a whole field of wells might produce as much as an ordinary power plant burning high sulfur fuel. In some aras this might require the use of a closed steam system.

In some cases, removing geothermal water can cause the ground to subside. The obvious solution is to reinject water. The oil industry already does this when necessary and for the same reason. It has the further advantage of replenishing the water needed to produce steam.

A question that remains unanswered is whether subsidence or a change in the underground heat balance might trigger earthquakes. Geothermal sources are located, after all, in earthquake-prone areas. So far there is no reason to suspect that any of the geothermal projects presently in operation have caused earthquakes. Also, even the most extensive operation can tap only a tiny percentage of the heat present in a geothermal resevoir.

The disadvantages of geothermal energy, however, are trivial when compared with the advantages. First, geothermal technology is simple. It does not require the complex and sophisticated technology that nuclear power does. Because of this, it can be developed much more rapidly. The cost of geothermal energy compares very favorably with fossil fuel plants. There is no fuel, no boiler. As costs of fossil fuels rise, geothermal energy will become even more attractive. Geothermal energy is clean. No fuel has to be mined, processed, or burned, and there are no wastes to dispose of.

If geothermal power is such an exciting possibility, why hasn't it been developed sooner? The United States has five and a third million acres of land that may have geothermal possibilities. The states of Arizona, California, Colorado, Idaho, Montana, Nevada, New Mexico, Oregon, Utah, Washington, and Wyoming are included. However most of these lands are federally owned, and until very recently they were closed to development. When the Department of the Interior opened bids for leases, it was astonished to recieve 2,456 applications. The granting of leases is the first step in the detailed exploration necessary to determine the full extent of our geothermal resources.

Once they have their leases, the lease holders will face a maze of overlapping federal, state, and local laws relating to gas, water, and mineral rights—for geothermal operations involve all three. Legal complexity has been another major barrier to geothermal development. New laws are needed to take the special problems of geothermal energy into account.

Overall, geothermal energy has a bright future. It might one day supply 10% of U.S. energy needs. It has been estimated that world geothermal resources amount to at least 30,000 MW. At present no more than 900 MW have been developed. Today El Salvador, Iceland, Italy, Japan, Mexico, New Zealand, the U.S. and the U.S.S.R. have geothermal power plants in operation. Chile, the French West Indies, the Philippines, Taiwan, and Turkey are all planning to build plants. At least twenty other nations are looking into geothermal power. As an easily developed source of cheap power it will be of great benefit to the developing nations. As a versatile source of clean power, it will benefit us all.

# Ocean Thermal Power

The greatest part of the solar energy that reaches the earth is received by the oceans, simply because they cover three quarters of the earth's surface. In the tropics the sun warms the surface of the ocean to a pleasant 25°C (77°F). Only the surface is warmed, however, and in the deeps the temperature stays at about 5°C (41°F). Because the cold water is denser and heavier than the warm water there is little mixing between them.

In theory any temperature difference can be used to run a heat engine and produce work. A boiler is placed in the hot region and a condenser in the cold region. The boiler vaporizes a fluid and the expanding vapor runs a turbine, producing mechanical work. Then the vapor is run through the condenser and cooled. It turns back into a liquid and is ready to go back to the boiler. This cycle, which is the basis of most power generation, is called the Rankine cycle. There is nothing to prevent the warm waters of the ocean from being used as the hot part of the cycle and the cold waters of the deeps for the cold part. Such a system could provide an almost unlimited supply of clean power. In effect it would be a solar energy system with the whole ocean serving as both collector and energy storage.

The idea of using the temperature differential in the ocean as a source of energy was first suggested by the French scientist d'Arsonval in 1881. He suggested several different situations in which a small temperature difference could be made to do work. One used a hot spring for heat and a river for cooling. Another used a river for heat and a glacier for cooling. In this way, d'Arsonval pointed out that the actual temperatures involved were not as important as the difference between them.

Others elaborated on these ideas but it was not until 1926 that they caught the attention of the man who was to put them into practice. Georges Claude was a French engineer who had developed several important industrial processes. With the income from his royalties he was well equipped to pursue his interest in generating power from the sea.

The first lab experiment was made in November, 1926. Two large flasks were connected together with a small turbine between them. One flask was filled with warm water at 82°F, the other with crushed ice at 32°F. While water normally boils at 212°F, it will boil at much lower temperatures if the pressure is reduced. Accordingly, a vacuum pump was connected to the system.

When the pressure was reduced to about 3% of normal atmospheric pressure, the warm water began to boil. Water vapor passed through the turbine and condensed upon the crushed ice in the second flask. A little generator connected to the spinning turbine produced enough current to light a few small light bulbs. Thus it was demonstrated that a temperature difference as small as 50°F could be used to generate powr.

Encouraged, Claude had a larger prototype built and installed at a steel mill in Ougrée, Belgium. The heat source was the warm water that had been used to cool the blast furnaces. The cool water was provided by the river Meuse. The unit went into operation in April, 1928. With a temperature difference of 36°F, it was able to generate 50 kilowatts.

The prototype also provided the answer to an important question. It contained a lot of auxiliary equipment, all of which consumed power. Some wondered if the system would be able to produce a net power output. In fact, the auxiliary equipment used only 25% of the total power generated.

The next step was to set the plant up by the sea so that it could use ocean water to generate power. The site had to meet three requirements: First, warm surface water. Second, a current to sweep the 'used' water away and prevent it from mixing with the intake water. Third, a steeply sloped bottom so that the cold water intake pipe could be as short as possible.

The site that Claude picked was Mantanza, Cuba, about 50 miles east of Havana. The surface water there was 82-83°F and there was a good current. The bottom, however, did not slope quite as steeply as Claude would have liked.

With the laying of the cold water intake pipe, Claude began to run into bad luck. The pipe, made in France, was to be sunk in sections and assembled by divers. The first section was laid without trouble, but a longer second section was lost in bad weather. Claude had a second pipe made from local materials. This pipe was to be floated out and laid in one piece, but it buckled in the center when the crew fouled up its instructions. The backers began to get restless, so Claude used his own resources to build and lay a third pipe. This time all went smoothly.

Claude's bad luck was not over, however. The pipe brought up water from 2000 feet down, but its temperature was only 58°F instead of the 40°F Claude had hoped for. The plant produced only 22 kw instead of 50. This was not even enough to run the auxiliary equipment, for Claude had installed an oversized cold water pump. He found himself having to buy power to keep the plant running. While the plant was not a success, it was not due to any defect in the basic principles involved. The plant did draw power from the ocean, and Claude hoped to solve its problems in his next project.

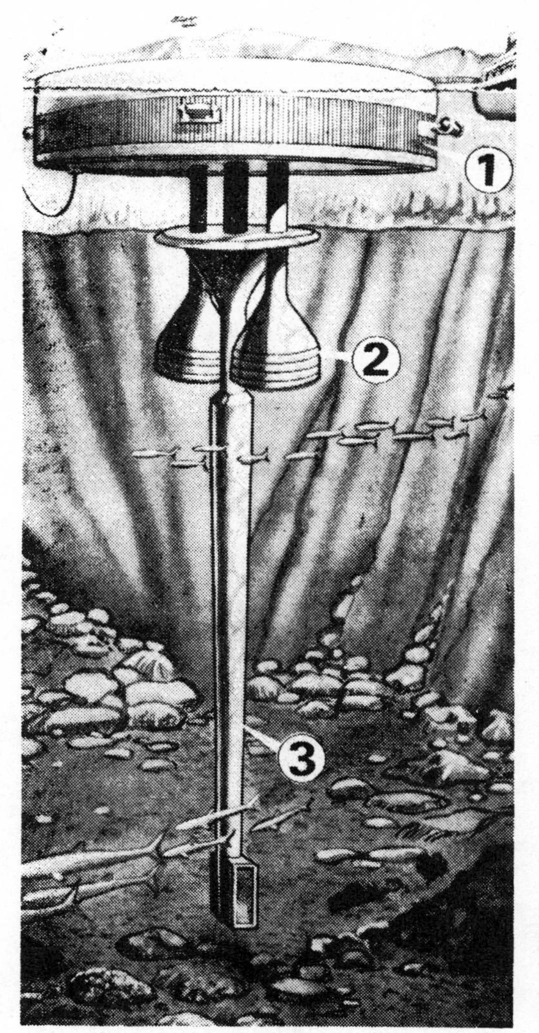

**A FLOATING GENERATOR system to produce power from differences in ocean temperature would include (1) a turbine generator turned by refrigerant gas heated and vaporized by warm surface water; (2) condensers where deep, cold water would liquefy gas for return to the generator and (3) a tube extending 1,800 feet down to draw up cold water for use in condensers.**

Giving up on shore plants, Claude mounted an 800 kw plant in the Tunisie, a 10,000 ton steamer. This time the cold water intake pipe was to be suspended from a float beside the ship. To use the power generated, Claude installed refrigerating equipment. He expected to make half a ton of ice a day at one fifth the usual cost. The ice was to be sold for use in air conditioning systems. Tests began in 1934. The equipment worked, but poorly, and it needed constant coaxing. Disgusted, Claude sank his equipment in the sea and returned to France.

A French society kept Claude's ideas alive. In 1948, the French government put up 4 million dollars for the construction of a 7,000 kw plant at Abidjan, on the coast of West Africa. It was hoped that the plant would supply energy as cheaply as a hydro-electric project. The plant was built and it ran, but problems with the cold water intake pipe kept it from being a success.

The sad part about Georges Claude's failure is that he might have succeeded if he had used a different system. Many of his difficulties stemmed from the fact that he insisted on using sea water as his working flud. This system is simple at first glance, but it has several drawbacks. The use of low pressure water vapor requires the use of huge turbines that are not very efficient. (The turbine at Abidjan was 26 feet in diameter and ran at 600 rpm.) Since the system must be operated at less than atmospheric pressure, power has to be diverted to run vacuum pumps and to remove dissolved air from the water. Last but not least, sea water is corrosive.

Several of Claude's colleagues suggested that he use an easily liquefied gas such as ammonia or sulfur dioxide in a closed cycle. The warm ocean water would circulate through a heat exchanger in which the working fluid would be vaporized. The vapor would spin the turbines and then go through a condenser cooled by cold sea water, where it would be turned back into a liquid to repeat the cycle. The vapor of the working fluid would be much denser than water vapor, so a smaller, more efficient turbine could be used. There would be no problems with corrosion or dissolved gases, and no need for vacuum pumps. Claude objected that the cost of the working fluid would make the system more expensive. The system would also be more complex, and heat losses in the boiler would reduce the low efficiency of the system even further.

Engineers J. Hilbert Anderson and James H. Anderson revived the idea of thermal ocean energy in 1964. Their experience with geothermal power plants had convinced them that with modern methods power could be generated from the oceans easily and inexpensively. They imagined a huge rectangular floating plant 40 feet wide and 360 feet long. It would use propane for its working fluid in a closed cycle. With a generating capacity of 100 MW it would produce energy at a cost of 3 mills per kilowatt-hour.

The Andersons' proposal aroused considerable interest. In 1972 the National Science Foundation funded a project to investigate the feasibility of the ocean power plant. A group of engineers and scientists headed by William Heronemus has been busy studying turbines, heat exchangers, hull construction, site characteristics, and ll the other details that would go into the operation of such a plant. Their conclusions are that such plants are technically feasible. Work could be begun on a prototype today with confidence that it would work as designed. No breakthroughs are necessary. In fact, nearly every part has a counterpart in some existing system.

Their preliminary design is for a 400 MW plant anchored in the Gulf Stream off Florida. The plant is designed with a site 25 km (16 mi.) east of Miami in mind. The poorest conditions for power generation occur in October, when the surface water is 80.6°F and the deep water is 49.2°F.

The Mark One would consist of two huge cylindrical concrete hulls, each 600 feet long and 100 feet in diameter. The Mark One would be a sizable job for a shipyard. As big as any two ordinary ships, its bulk would be exceeded only by the larger oil tankers. It would spend most of its time below the surface, where it would be almost immune to wind and wave action. The twin hulls would be surmounted by four large flat heat exchangers, and the depth of the Mark One would be carefully adjusted to keep them just below the surface, where the water is warmest. With its ballast and trim tanks, its many compartments, and its watertight doors, the interior of the Mark One would resemble that of a submarine. Its crew would be carefully trained in basic U.S. Navy submarine safety practices.

Inside the two hulls would be 16 turbo generator sets, each one capable of an output of 37.5 megawatts. They would be big, built for maximum efficiency rather than compactness, and the whole structure of the Mark One would be designed around them. The working fluid would probably be propane. Ammonia has also been considered because it allows a more compact and efficient system; but it has the drawbacks of being toxic, irritating, and corrosive.

Also in the hulls would be the condensers. Cold water for them would be brought up from the depths at the rate of 30 million gallons per minute through a pipe 80 feet in diameter and 1,100 feet long. The water would flow so fast that it would have no chance to warm up during its trip through the uninsulated pipe.

The pipe would also be part of the mooring system. It would contain ballast compartments along its length and at its other end there would be a concrete barge 300 feet long, 60 feet wide, and 30 feet deep. The barge would also be divided into ballast tanks. When complete the entire system would be towed out to its site as a unit. There ballast tanks would be slowly and carefully flooded to put pipe and barge exactly into position. Once in place, the barge would serve as a great anchor.

The completed system would consist of many plants spread out over a 15 to 550 mile area of the Gulf Steam. They would send their power ashore through underwater cables, or, if it proved more economical, use their power to break water down into hydrogen and oxygen which would be sent ashore by pipeline.

The goal of the study group is to design a system that would be capable of supplying at least 10% of the U.S. energy demand. At the same time they want to make the ocean thermal power plant a more attractive investment than either nuclear or fossil-fueled plants. The original target was a system that would cost no more than $600 per kilowatt to build. While the economic analysis is not yet complete, indications are that the ocean plant could cost $1,100 per kilowatt in 1980 and still compete with fossil fuel or nuclear plants. The Mark One is only a preliminary design intended to point the way towards more refined models.

A system of ocean thermal power plants could provide many other benefits. They could be designed to produce fresh water in addition to power. Hydrogen and oxygen could be produced by electrolysis. With abundant power available, the power plants could become nuclei for floating industrial plants that would extract minerals from sea water. The fishing industry would also benefit. The cold waters of the deeps are rich in nutrients. When brought to the surface these nutrients become available to algae and other small organisms. Their numbers would increase, in turn providing food for a greater number of fish.

The construction of a system of floating ocean thermal power plants would be an ambitious and expensive undertaking. In many ways, however, it is a far more attractive alternative than nuclear power. It doesn't involve anything we don't know how to do right now, and it would be an inexhaustable source of clean, safe, dependable power.

# Wood Heat

By Gary Wayne

I guess that I have tried most of the popular ways, or should I say, the most commonly used methods of heating a home. I've used coal, and discovered that it's dirty. Oil is cleaner, but not by much. And besides it requires a fan and other paraphernalia which I have never quite succeeded in understanding well, much less to operate well. Natural gas is very clean and cheap, but you must be close to a service line, and the stuff scares me. It can asphyxiate its victims, or failing that, it can blow them to kingdom come. Electric heat is expensive, and it strikes me as being too artificial. Another thing that I don't like about electric heat is that it doesn't move the air as do other forms of heat that require combustion. That leaves me with wood heat. Wood heat is simple. Here in Washington State, wood is cheap and easily available. I like the smell of wood heat and I like the looks of wood heat. When I come inside from the cold out of doors, I like to get up close to the fire and get warm fast, and then I start backing away from it until I find a location with the proper temperature. I grew up using wood heat and its warmth is familiar to me and I feel confident in its use.

Obtaining wood for fuel is not a problem, since there are so many ways in which it can be obtained. If a guy has a pick-up truck, an axe and a chain saw, he not only has the means by which he can obtain his own fuel, but also he can have fuel to sell for extra income.

Where big logging companies have clear cut a forest area, there is an incredible amount of waste left on the ground that is quite suitable for fuel. One has only to get permission and then help himself. I have never had much trouble in getting permission, because most people are glad to have someone clean up the mess.

Sometimes a timber or pulp company will buy up a track of timber that partially contains some types of trees that are of no use to them. Such as alder, which makes an especially good fuel. If you talk to the right people, they would consider it a favor for you to go in and get that alder out of their way.

Watch for houses to be torn down. This can get you fuel and lumber to sell or use. Look for construction sites. They waste a lot of odd ends and pieces of lumber.

If one is fortunate enough to own some acreage, he may find it convenient to grow a wood lot. Five acres should be enough to provide small amounts of timber and sufficient fuel to heat a reasonably well insulated house in the winters in the continental U.S., if it's not too large or drafty.

One should consider his needs, the availability and the conditions for the types of trees that he is going to plant. Western hemlock and cottonwood are of little use and should be cut from the wood lot right away. Broad leaf trees provide a fuel that burns slowly and hotter than do the evergreens. Oak makes an especially hot and long lasting fire. Alder is well suited to wet and poorly drained soil. Pine does well where there is little top soil. One would be well advised, however, to plant a wide variety of trees so that if a blight or pest strikes, he will not lose the entire lot.

When planting conifers, plant them close together. In a few years you can thin them out, and sell the cut ones for Christmas trees. As the trees get larger you can continue to thin them out, using them for beams, poles and fuel. It is well to cut the lower limbs and keep them crowded so that they tend to grow the

# Thin out by cutting every other tree....

green stuff at the top and leave the trunk free of knots. Thin out by cutting every other tree.

If you cut your wood green, or should I say, live, be sure that you give it time enough to cure or dry out. About 2 or 3 months in a dry warm place should be plenty of time. If you attempt to burn wood green, it will burn slowly and make a lot of smoke and cause you problems with soot in the flue. I can tell you from experience that getting through the winter on green wood is a real drag.

If you decide to use a wood cook stove, be sure to place it in the kitchen so that you can step away from it once in a while. If the stove is set up close to the sink or something, you will find that doing the dishes on a hot summer day is real hot work. A kitchen with a wood stove in the summer is much like a sauna unless you manage to set it up somewhere that is well ventilated and out of the way.

The kind of stove that you use for a space heater will depend on several factors. Like: practicality, expense, asthetic appeal, availability, etc. The most efficient way to squeeze the BTU's out of the fuel is to use a stove that will close up tightly. That way you can control the draft and air intake in a way that will give the desired heat with a minimum of bother on your part. The best of the closed stoves that I know of are the Rightway and the Ashley. They circulate the hot gases around the combustion chamber in such a way that they are ignited a second time, thus insuring complete combustion of the fuel. Some even have a thermostat on it so that you can set it to a desired heat and the thermostat will regulate the amount of air entering the stove and thus controlling the size of the flame. Another nice feature about these stoves is that they require stoking only once a day.

Next there is the metal fireplace to consider. Among these the Franklin is my own personal choice. With the Franklin stove I have heat radiating in all directions from the stove to fill the space equally on all sides with heat. I can close it up at night with a fire going, and when the fuel is exhausted and the fire out, I don't lose heat through the chimney as I would with an open fireplace. The Franklin has the advantages of both the closed stove and the open fire. I appreciate an open fire. There is a charm and a magic about it that is hard to put into words. To see an open fire and to be close to it appeal to instincts that are much older than us. It's an instinct developed a long time ago when a fire meant warmth and security from prowling carniverous animals. The sight and feel of the open fire can be had with the open fireplace also; however, the open fireplace doesn't have the extras of the Franklin stove.

If one is going to have an open fireplace, I would advise him to get a free standing metal one, because it radiates the heat better than one made of masonry. If one needs a fireplace that fits into a wall, then I would recommend a Heat-a-lator. This is a metal fireplace in the wall with masonry around its outside and with an air space between the metal and the masonry. There are also vents so that the air can be moved around the metal and thus circulate the heat around the room.

If one is going to build a fireplace of masonry, there are 2 good books that I could recommend, namely, "The Owner Built Home" by Ken Kern, and "The Foxfire Book." Both of these books seem to treat the subject well. I have never built one myself, and therefore feel unqualified to pursue the subject any further.

If you are going to have a wood fire inside your house, it would be wise to have an air inlet so that fresh air can get to the fire. Most people are reluctant to bring cold air inside, because that is just like defeating the purpose of the fire. However, for a fire to burn well it must have air. If not provided, it will draw from cracks in the door or windows thus creating drafts. Or worse, it might create a vacuum in the house which could prevent smoke rising through the chimney.

A good deal of care should be used in setting up the chimney. As a matter of fact, this might be the most frequent oversight that causes failure in a wood heat system.

As the metal chimney leaves the stove, it carries a lot of heat with it. It behooves us, therefore, to get as much exposure from the chimney as possible before it passes outside and dissipates its heat out there. If the chimney is to pass through the wall, pass it through near the ceiling rather than near the floor. Better yet, pass it through the ceiling if you can. Some people place a small fan behind the chimney; and there is a fixture available that places a fan in the chimney so that the air can be circulated around the hot metal and pull the heat off it before it goes outside.

Now I would like to point out what I have observed as the most common cause of fire caused by a newly set-up system of wood heat. When the chimney passes through the wall or ceiling, be sure that it is well insulated and that no part of the hot metal touches, or is anywhere near, the wood framing or siding of the house. This metal gets hot and I have seen it start fires more than once.

When the chimney goes through the wall or roof, it should in some way be insulated so that the smoke will not turn cold before it gets all the way out of the stack. Cold smoke does not rise as will hot. A masonry chimney does not seem to present this problem; however a metal one does. If a metal chimney is used, a metalbestos or insulated chimney should be used in the attic or above the roof line.

The chimney should rise above the highest part of the roof. This is so that the wind can pass over the top edge of the stack unimpeded. Those of you who have studied physics will know that this causes an updraft in the chimney. This is known as Bernoulli's Principle, and is applied in many ways, such as vaporizers, perfume bottles and hand sprayers.

Using the principles set down here, one should be able to install a system of wood heat and enjoy the many comforts of this most versatile fuel. Good luck to you.

Gary W. Smith
1900 Valencia St.
Bellingham, WA 98225

# Tidal Power

No one can spend very much time by the shore of the sea without becoming conscious of the twice daily rise and fall of the waters we call the tides. There are two high tides and two low tides each day. The tides do not recur at the same time each day, however. Today's high tide takes place about 50 minutes later than yesterday's. A complete cycle of two high tides and two low tides takes 24 hours and 50 minutes. This is sometimes called the "tidal day." The result is that the times of high and low tide shift around the clock.

The extra 50 minutes comes about because the tides are due to the gravitational pull of the moon. As the earth turns, the moon moves in its orbit, and in one revolution of the earth, the moon has moved enough to delay the tides by 50 minutes.

The gravitational pull of the sun also influences the tides. Because the sun is so much farther away than the moon, its influence is only about one third as great. This is enough, however, to cause an important variation in the level of the tide. When sun, moon, and earth are in a straight line, as during new or full moon, the sun and the moon pull together to cause the highest high tides and the lowest low tides. These are called spring tides. During the moon's first and last quarter, the sun and the moon are at right angles, and their pulls are working against each other. The result is the neap tides.

The difference between high and low tides is called the range of the tide. The spring tides have the highest range, typically about twice that of the neap tides.

In the open sea, tides amount to only one or two feet, but where they are funneled by a bay or river mouth, the total rise and fall can be 10 feet or more. What could be simpler than to throw a dam across a cove or inlet and let the daily rise and fall turn a mill wheel? Tidal mills of this sort, very similar to river mills, were used on the coasts of England and France as early as the 11th century. Many were built by the American Colonists, and some of these mills were used well into the 19th century.

The development of the steam engine caused most of the early tidal mills to be abandoned. Today the need for new sources of clean energy is causing engineers to take a new look at tidal power.

The simplest way to harness the power of the tides is to allow the rising tide to fill the basin behind a dam, close the gates at high tide, and as the tide falls allow the impounded water to pass through turbines on its way back to the sea. Because this method uses only the ebb of the tide, it makes available only about half the potential tidal energy. If the tide is made to turn turbines during the flood as well as the ebb, about half again as much energy can be realized. To do this, the sea is allowed to pass through the turbines into the basin as the tide rises. As high tide approaches, the difference in levels (called the head) becomes too small to operate the turbines, so they are stopped; sluice gates are opened, and the sea is allowed to fill the basin as rapidly as possible. When the levels of basin and sea coincide, the sluices are closed and everything waits while the tide drops. When there is once again a large enough head, the basin is allowed to drain through the turbines. As low tide approaches, the head again becomes too small to drive the turbines, so the sluice gates are opened and the basin is allowed to empty as rapidly as possible. When the basin has reached its lowest level, the sluices are closed, and everything waits while the tide rises. When there is once again a sufficient head, the sea is allowed to pass through the turbines, filling the basin and beginning a new cycle. Because it is impossible to fill or empty the basin instantaneously at the turn of the tide this method cannot capture 100% of the power available in the tide.

The main drawback in both of these systems is that they are intermittent. They generate no power during slack water. Also their schedule of power delivery is based on the tidal day of 24 hours and 50 minutes, so that each power period occurs 50 minutes later than it did the day before. Clearly this is highly inconvenient unless there is some means of storing the power, or unless an industry can adapt itself to this peculiar schedule.

To overcome this there are schemes for continuous power generation, using two or three basins instead of one. One basin is emptied at low tide, and one basin is filled at high tide, and the flow of water through the turbines is shifted back and forth between the basins and the sea so that there is always a reasonably constant head available. The disadvantage of the continuous systems is that they offer only about half the power available from the intermittent systems, and they are more complicated to build and operate.

There are some disadvantages that are common to all tidal plants. One lies in the variation in range between spring and neap tides. If a plant is built to utilize all the power in the spring tides, then a large part of its capacity will be idle during the neap tides. If a plant is built to handle only the range of the neap tides, then the extra power in the spring tides is wasted. Another problem stems from the fact that a tidal power plant is a large scale project, affecting a wide region. Damming a large body of water to turn it into a tidal basin has strong effects on fishing, navigation, drainage patterns, and so forth, and it is difficult to satisfy all concerned.

In spite of all these difficulties there are two large tidal plants in operation today. One is the Rance estuary plant, which has been working on the French coast since 1967. The other is a 1,000 KW experimental plant built by the Russians at Kislogubsk on the White Sea.

# The Rance Tidal Power Plant

In the middle 50's French engineers began design studies for what was to become the first major tidal power project in the world. At the time nuclear power was prohibitively expensive and complicated. The Suez crisis gave added impetus to the project by underlining Europe's dependence on imported power. The location selected was the Rance estuary, which opens into the Gulf of St. Malo. Here are found the highest tides in Europe—a range of 9 to 14 meters (30-46 feet) is typical. During the highest tides water flows in or out of the estuary at the rate of 630,000 cubic feet per second. Schemes for harnessing the energy of these tides go back to 1909.

The very tidal current for which the site was selected made construction difficult. After extensive model studies work was begun in January, 1961. A double row of cofferdams was sunk across the estuary and the water pumped out from between them. On the dry bed a dam 710 meters (2,329 feet) long was built. Unlike most dams, this one had to be designed to withstand pressure from either side. When completed the dam formed a basin of 22 square kilometers, with a useful volume of 184 million cubic meters.

Included in the dam is the powerhouse, which is 1,200 feet long. Running the length of the powerhouse is a long gallery which contains cranes and machinery for servicing the turbogenerator units below. The turbogenerators are the heart of the installation. There are twenty-four of them, each rated at 10,000 KW power output. Each one is suspended in a tapering conduit 174 feet long. A streamlined bulb houses the actual generator, which is turned by a turbine that looks like a giant ship's propellor. Each turbine is 17.5 feet in diameter, and the angle of the blades can be changed according to the head and rate of flow to keep the generators turning at a steady 94 rpm. The first four units were put into service in July, 1966. The entire plant was completed in 1967 at a cost of roughly 100 million dollars.

The reversible turbines at La Rance

allow power to be generated on the flood as well as on the ebb. Usually about 8 times more power is generated on the ebb than on the flood. This is not the end of the story, though, for the design of the turbogenerators permits an interesting trick which can be used to increase power output even further.

We start at slack water with the basin nearly empty and the tide rising. When the difference in levels is enough to run the turbines, water flows from the sea through the turbines filling the basin. The level in the basin rises until the head is no longer enough to run the turbines. Usually at this point the turbines would stop and the plant would sit idle waiting for the tide to fall. Instead, outside power is fed to the generators. This causes them to act like motors, turning the turbines, which pump water into the basin. Later, when the tide has dropped enough, this extra water will run through the turbines to generate power.

The same trick can be used on the other half of the cycle to pump water out of the basin. If this sounds rather pointless, like taking a dollar out of one pocket and putting it in another, consider that the pumping is done against a very low head, and the generating is done at a much higher head. The extra output more than makes up for the energy used to do the pumping. In fact pumping increases the power output of La Rance by about 25%.

Another point to consider is that power is worth more at times of the day when demand is heaviest. By doing more pumping when power is cheap, La Rance can produce more power during hours of peak demand. To get the most out of the plant, the level of the basin, the level of the tide, the time of day, and the anticipated power demand all have to be considered together. At La Rance this is done by a computer and the large sluice gates are in almost constant motion.

The successful operation of La Rance has proven that tidal power is both practical and dependable. In other nations tidal plans long-shelved are being re-examined with renewed enthusiasm. While tidal power is far from being a total solution to anyone's power needs, it certainly deserves a place in our plans for the future.


**References - Tidal**

**The Tides**
Edward P. Clancy
Doubleday, 1968.

"Tidal Power Comes to France"
**Engineering,** July 1, 1966
p. 17

"Tidal Energy
Roger Henri Charlier
**Sea Frontiers,** vol. 15, 1969, p. 339.

**References - Geothermal**

"Geothermal Power"
Joseph Barnes
**Scientific American,** January, 1972, vol. 226, no. 1, p. 70.

"Geothermal Energy: An Emerging Major Resource"
Allen L. Hammond
**Science,** September 15, 1975, vol. 177, p. 978

"Dry Geothermal Wells: Promising Experimental Results"
Allen L. Hammond
**Science,** October 5, 1973, vol. 182, p. 43.

"Geothermal Resources"
Eric Burgess
**Analog,** May 1974, vol. XCIII, no. 3, p. 56.

**References - Ocean Thermal Power**

"Power From the Tropical Seas"
George Claude
**Mechanical Engineering,** December 1930, vol. 52, no. 12, p. 1039.

"Ocean Temperature Gradients: Solar Power From the Sea"
William D. Metz
**Science,** June 22, 1973, vol. 180, p. 1266.

**Research Applied to Ocean Sites Power Plants**
J. Hilbert Anderson
NSF/RA-N-74-002 January 1974.


# Tidal Grist Mill

This mill is typical of many that were built and operated from colonial times up to the Industrial Revolution. Appropriately enough, it is located in Tidewater, Virginia.

The original mill was built in the 1760's. An earth dike was built across a cove, closing it completely except for a sluice. There the ebb and flood of the tide turned a water wheel which powered a mill little different from the water mills of the time.

During the Revolutionary War, the mill ground grain for the Continental Army. It was burned during the Civil War and later rebuilt. Old-timers recall that the mill was in full operation right up to 1912.

The local tides are about 34 inches. To operate, the mill needed a head of only 14 inches. It could mill 32 bushels of grain per tide. The present owner estimates that it produced less than 5 horsepower. Today a concrete plug blocks the sluice, and boats (and tides) run freely in and out of the cove through a wide breach in the old dike.

(Thanks to Eric P. Johnson)

*Dear Sirs:*

*Enclosed is a copy of a story written for "Alternate Sources of Energy" which gives a brief history of our project, and a copy of SAE paper #719046 which we presented in Boston at the 1971 Intersociety Energy Conversion Engineering Conference, the only publications we have circulated.*

*Due to the large volume of responses from "Production", I will answer all your questions in this one letter and also anticipate some that were not asked.*

*When we started this project in December, 1969, we thought there would be no problem of getting funding or general interest in Hydrogen as the only clean fuel, but it took the energy crisis to generate enough interest so that our press releases didn't get dumped in the nearest circular file.*

*All our primary research and testing was done in 1970 and the F-250 pickup truck has been parked since then, except for a brief operation in mid 1973. Due to lack of funds, which have almost exclusively been supplied by ourselves, the period of 1971 to the present has been spent on obtaining patent coverage. Our patents are still pending.*

*Besides presenting the SAE paper in August, 1971, we then also visited Detroit Diesel, American Motors, Ford, & Chrysler where we presented our case for the $H_2/O_2$ car to the respective engineering departments.*

*In a letter to us, the EPA stated they are only interested in funding Brayton cycle, Rankine cycle, & stratified charge engines which run on gasoline. (Also see SAE paper #729125 by Mr. Brogan)*

*So, for all practical purposes, we have been unable to make any significant progress in the last 3½ years, and thus cannot yet compare the performance of our system, which was originally intended to be a retrofit package to make every existing gasoline fed engine into a non polluting one, with existing engines. There are still bugs in the system, which can only be chased out with expensive instrumentation. We have every reason to believe that, contrary to engines which use $H_2$ & air, the performance will exceed that of engines fueled with gasoline & air, and, with the on board separation of $O_2$ from the air through the use of the cold liquid $H_2$ as a means to liquify the air prior to separation, this sytem will be safer than the use of gasoline.*

*Note that the engine uses $H_2$ & $O_2$ in gaseous form only and stores the gases in liquid form. Also, due to the higher efficiency of $H_2/O_2$ combustion vs gasoline, approximately the same amount of water vapor is exhausted per mile for each system.*

*We have not yet incorporated nor formed a partnership as we felt the money to do so would be better spent on our research.*

*Thank you to those of you who sent stamps to help pay the cost of mailing.*

*Sincerely,*
*Paul Dieges*
*714-657-2822*
*for The Perris Smogless Automobile Association*

# The Hydrogen-Oxygen Car

From Alternative Sources of Energy #11

**In January, 1970, Ben Minnich, Fred Nardecchia, Pat Underwood, and myself made our first primitive attempt to operate an engine on hydrogen and oxygen.**

Our first effort was to get a junked car with an engine that still ran. We planned to operate it using $H_2$ and $O_2$, as opposed to $H_2$ and air mentioned in the A.S.E. issue.

Our theory was that the standard auto engine was basically well-designed, but that something should be done to its fuels to enable it to be operated free of pollution. We knew that $H_2$ and air would work—but as with all conventional fuels when burned with air, nitrous oxides (NOx) are formed and thus the engine would not be pollution-free.

We also knew, as does any high school student, that 2 parts $H_2 > 1$ part $O_2$ makes 2 parts $H_2O$ (water). We also knew that the Hindenburg dirigible was filled with $H_2$ when it caught fire and burned, frightening many aboard to jump to their deaths, instead of riding it to the ground and escaping relatively unharmed.

Accordingly, we conducted our experiment by remote control, with us on one side of a huge boulder and the car on the other side. After much popping and banging of the engine, we determined that the mixture ratio that enabled the engine to operate best was not 2 parts $H_2$ to 1 part $O_2$, but more like 8 parts $H_2$ to 1 part $O_2$, by volume. This is necessary as there is no nitrogen ($N_2$) from the air being used as a dilutant, only pure $O_2$.

To save and re-use this excess of $H_2$, we devised a method of separating the water in the exhaust from the unused $H_2$, and recirculating it back to the engine intake for reuse, a concept which forms part of a patent that is pending on our invention.

The first vehicle to operate over the highways using our new system was my 1930 Ford Model A pickup truck. The historic date was Feb. 21, 1970, when the truck first moved under its own power from $H_2$ and $O_2$.

As a result of testing, we determined that we were getting an economy of operation of 10 miles per pound of $H_2$ as compared to 2½ miles per pound of gasoline (about 15 m.p.g.) when the same course was traveled. This four-fold increase was not expected, as $H_2$ only has 2½ times the chemical energy as gasoline. It turns out that $H_2$ burns with so little radiant heat (about ⅓ that of gasoline) that very little energy is lost into the cooling system through the engine block, and accordingly more of the fuel is available for useful work.

The disadvantage of an $H_2$-$O_2$ fuel system is that for every pound of $H_2$ carried, eight pounds of $O_2$ must also be carried. The disadvantage of using compressed gas storage, as we did with the Model A, is that one 150 lb. cylinder holds only one lb. of $H_2$ compressed to 2,000 pounds per square inch when full, enough for 10 minutes of running at 60 MPH. Obviously, it is necessary to use other methods of storage in the real world.

The method we feel is best to use for vehicles is to store the $H_2$ and $O_2$ as a liquid. When $H_2$ is cooled to near absolute zero, it becomes a liquid, the lightest liquid known, weighing only 23½ pounds (#) for 40 gallons. 23½ # of water would occupy under 3 gallons.

$O_2$, when cooled to about halfway between absolute zero and the freezing point of water, becomes a liquid also. 188 # of $O_2$ occupying 20 gallons of volume, is the amount necessary for combustion of 40 gallons of $H_2$.

Thus, using 10 miles per # of $H_2$ (some using other systems claim as much as 20 miles per #) as a basis, the range of the vehicle using 23½ # of $H_2 >$ 188 # of $O_2$ (a total of 211½ # of fuel occupying 60 gallons of volume plus the insulated tanks) would be 235 miles to perhaps 470 miles.

We installed a liquid storage system which stores 47 # $H_2$ and 376 # $O_2$ on a 1960 Ford pickup in anticipation of running the vehicle in the 1970 clean air car race, but were not able to complete the system in time. Lack of financing has prevented us from getting the system perfected, although the truck does run today.

Our future work will be oriented to developing a method of separating the $O_2$ from the air as the vehicle moves down the road, so the weight of $O_2$ will not have to be carried.

In order to avoid carrying this weight of $O_2$, other experimenters have used a very lean mixture of $H_2$ and air to reduce the temperature of combustion and thus reduce the formation of NOx to levels below the 1976 (now 1977, due to the auto companies' wailing) air standards. However, it then becomes necessary to supercharge the engine in order to achieve power comparable to gasoline. At the proper mixture ratio, for the most efficient combustion of $H_2$ and air, the NOx production exceeds that of gasoline and air. Of course there are none of the other pollutants that gasoline puts out.

A very lean mixture can be used due to the wide range of flammability limits. A lean mixture of gasoline and air will not ignite unless something considerably hotter than a spark plug is available to ignite the mixture. When you hold a hose of low pressure, constant flow $H_2$ into the carburetor (dry of other fuels), you will find that the throttle can be opened and closed over a wide range to vary the mixture ratio and the engine will still run. When it runs fastest, that's when the most efficient combustion is taking place. If you then want to compare this action to propane or another gas, you will find that you can vary the throttle only over a rather narrow range. If you use too small a flow of $H_2$ or propane, the engine will backfire at you, so keep your face away.

I believe the salvation of this environment lies in the use of this synthetic fuel system of $H_2$ and $O_2$ wherein solar power (wind, tide, water, etc.) is transformed into electricity which is then used to break water into its components, which can be stored indefinitely, as opposed to an automotive-type storage battery. These components can then be used when needed in either an engine, using the system described above, or in a heating-cooling system for a house. Since there is no carbon in $H_2$ as there is in methane ($CH_4$), the flame can be used inside a dwelling with no vent other than for fresh air, as there is no way the poisonous carbon monoxide (CO) could be formed. This results in a more efficient use of the energy in the $H_2$.

Good luck, everybody, with your experimenting, and let me know how you do.

*Paul Dieges*
*Perris Smogless*
*Automobile Assoc.*
*PO Box 892*
*Perris, CA 92370*
*Ph: 714-657-2822*

# Hydrogen and Oxygen Combustion For Pollution-Free Operation of Existing Standard Automotive Engines

**By Patrick Underwood and Paul Dieges**
**Perris Smogless Automobile Association**

This paper has been prepared to circulate the result of some early approaches recently made to the possibilities of using the combustion of hydrogen and oxygen in standard formerly gasoline engines. The sole exhaust product, being water, the experiments were undertaken to exploit the obvious clean energy conversion potentialities. Stoichiometric combustion with air or oxygen did not seem to be practical with existing engine designs during the experiments and, indeed, subsequent research[1 2 3 4 5 *] indicates that even major design departures do not appear fruitful for reciprocating engines operating stoichiometrically.

***(Editor's note: Stoichiometric combustion means that the gases are combined in a ratio of two volumes of hydrogen to one volume of oxygen so that both are completely consumed with none of either left over.)***

And though hydrogen and air do not produce more NOx than gasoline at stoichiometric combustion, increasing the fuel to air ratio does not lower the NOx levels as fast as with gasoline, and such increase wastes a significant portion of the hydrogen into the exhaust[6]. Consequently it was determined that combustion characteristics would be useful with extremely rich fuel to oxygen ratios and, using a condenser and mechanical separator to remove the exhaust water vapor, a simple system to recover and reuse the unburned hydrogen was invented[10]. From outside the recirculation loop, the engine operates stoichiometrically. The engine heat loss formerly removed via the exhaust system is now added to the normal water jacket heat loss for release by the radiator. However, by revaporizing the collected exhaust water with some of the waste exhaust heat and by vaporizing the cryogenic substances, with the coolant, the cooling load is significantly reduced and the thermodynamics of the system are thus improved. In order to match the partial pressure of oxygen to that which occurs in air so that lubrication oxidation would not become a problem and so provide a chemically clean replacement for the nitrogen in air, as a working fluid, it was decided to operate the engine in a hydrogen-rich environment and to add oxygen in a manner analogous to adding gasoline to the air induction system on ordinary gasoline engines without the phase change therein. Thus the oxygen is present only from the mixing device to the combustion chamber. Although an effort was made to insure that the recirculating gas was hydrogen, no specific analysis was performed to see what miscellaneous gases were actually recirculating in the system. While complications due to inexperience with the gases and insufficient instrumentation arose during the experiments, several advantageous circumstances arose. Starting appeared to be easier and the closed nature of the system sharply reduced operating noise. Greater thermodynamic efficiency of the conversion of heat to mechanical energy due to the low flame radiance and noncondensability of the combustion gases appears to be the reason for getting more miles per pound of fuel than would otherwise be expected on the basis of the energy contents of hydrogen and gasoline. A contemplated but unverified advantage is reduced lubricant contamination as hydrogen burns cleaner than natural gas or propane.

## STUDEBAKER

The initial experiments involved attempting to operate a six-cylinder water cooled engine (Studebaker 1950) directly from pressure supplied hydrogen and oxygen force fed via the normal intake manifold at the carbureter mounting flange. Pressures on the order of 30 pounds per square inch were applied. The oxygen was admitted first with the intent being to standardize a pure oxygen flow to be followed up with a gradual build up of hydrogen flow with a corresponding reduction in oxygen flow until engine run was attained. Cranking was to be continuous during this experiment phase. The hydrocarbon residuals; oil, gasoline, and carbon deposits ignited prior to admission of any hydrogen and ultimately caused abandonment of this particular build up technique. This burning caused superficial engine damage and the procedure was reversed. Hydrogen standardization was easily achieved, and although some incidents of external combustion, due to leakage and collection of gas in the tail pipe and crankcase, ensu ed, no steady engine operation was achieved when the oxygen flow was increased. Experiments were then performed using hydrogen and air mixtures at various source pressures for the hydrogen without any sophisticated mixing system. The engine was thus successfully started when pressures of 1 to 5 p.s.i. were used. Having achieved a measure of control of the hydrogen and air operation, the air was then gradually enriched with oxygen until only pure oxygen was being added to the hydrogen. At this point, the supply pressures were on the order of 5 pounds per square inch at the point of measurement at the source end of fifty feet of ¼-inch rubber lines. Stable and controllable operation was developed when the hydrogen flow rate was consistently in excess of stoichiometric compared to the oxygen flow rate. The engine was tested to destruction by accidentally increasing rather than decreasing the oxygen flow. Observations made included smoothness, quietness and easily starting once a procedure was established.

The severe excess of hydrogen flow requirement led to the invention of the aforesaid system to recover the unburned hydrogen from the exhaust and to recirculate it back to the intake, there to be mixed with a stoichiometric supply of new hydrogen and oxygen.

## MODEL "A"

The recirculation system was then developed in a road qualified vehicle (Ford, Model A 1930). See Figures 1 and 2. The engine exhaust was passed through an old automotive radiator immersed in a water filled, 55-gallon drum connected to the engine cooling system to condense the water. The hydrogen was then separated from the water in a centrifugal separator made from a piece of 4″ PVC pipe for the aforesaid reuse. The mixing device was no more than the junction of 4 pipes and a connection to the intake manifold. The hydrogen and oxygen supply pressures to the control valves ahead of the inlet to the mixing device were on the order of 10-20 pounds per square inch. The oxygen source pressure was tracked to the hydrogen source pressure at approximately twice the absolute hydrogen source pressure. Flow impedances for gases vary about as the square root of their molecular weight. Since the volumetric flow rate of hydrogen is twice that of oxygen and since the molecular weights of hydrogen and oxygen are in the ratio of one to 16, the one to two pressure ratio into similar input impedances is consistant with the observed requirement.

Fig. 1 — Detail of right side of 1930 Ford Model "A" engine showing mixing chamber, re-circulating pipe, connection to breather to collect blowby, and connections to hydrogen and oxygen supply. ¼″ tubing was connected to pressure/vacuum gauges

Fig. 2 — Right side of 1930 Ford Model "A" showing centrifugal separator, re-circulating pipe to engine intake, and exhaust pipe of ¼″ tubing, shown here delivering water to drinking glass, to measure flow output.

This system was operated over roads, data taken, and observations made. The vehicle had somewhat reduced acceleration, but could be made to achieve its former top speed. The failure to achieve more acceleration performance was probably the result of source impedance limitations. The 50-foot, one-quarter inch lines were still being used.

Several closed course runs over public roads were made using hydrogen and oxygen to get operating economy data. Computing the volume of gas used from observations of the cylinder pressure gauge and measuring the distance traveled, an economy of 10 miles per pound of hydrogen was observed to be required with the hydrogen and oxygen fueling. By duplicating the acceleration and other performance with normal gasoline operation, it was found, by measuring the volume of fuel used and the distance traveled, that 2.5 miles per pound of gasoline (about 14 miles per gallon) were required with gasoline and air fueling. This indicates increased thermodynamic efficiency since only 2.5 to 3.0 times the combustion energy is expected with hydrogen and oxygen compared with gasoline and air combustion. While some other bias such as poor carburetor performance may exist, the aforesaid two possible mechanisms of low radiance and low condensing temperature, exist that can account for this thermodynamic efficiency gain. The radiance of the hydrogen and oxygen flame is notably low, resulting in lower losses to the engine components. Neither hydrogen nor oxygen are condensable at engine operating conditions so that no phase change loss would occur nor could burning droplets on the cylinder walls conduct heat into the engine components during combustion.

In this vehicle, the exhaust water was continuously accumulated or allowed to drain off as a liquid. As a liquid, when cooled sufficiently, the water was found to be potable but somewhat oily and rusty tasting.

This first vehicle operation was made with compressed gas sources but it was intended eventually to use cryogenic gas sources. Since it is quite desirable to utilize waste engine heat to vaporize the cryogenic substances and to temper the gases, experiments related to heat exchanger requirements, using liquid nitrogen as a model, were made on this vehicle.

## PICKUP TRUCK

A more modern vehicle (Ford, F250 pickup, 1960) was later converted, using a more compact and sophisticated system. See Fig. 3. This vehicle ultimately was operated from cryogenic supplies of hydrogen and oxygen although some early operation was made with compressed gas sources. The cryogenic supplies of hydrogen and oxygen were carried in the pickup bed in shipping containers designed for transporting them and ultimately delivering the product at the fixed destination. To withdraw the product en route was of no significant technical departure from their normal application, although a substantial body of United States Federal regulations unintentionally inveighs against the practice from a legal viewpoint. There are provisions, however, exempting the

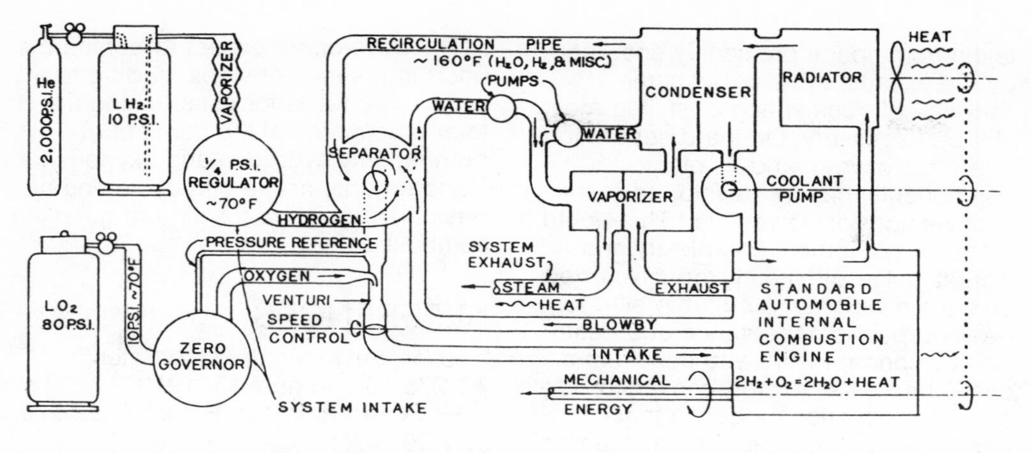


Fig. 3 — Flow diagram of Ford F-250 pickup

propulsion fuel supply from outright prohibitions.[11] Two hydrogen containers were provided to match the single oxygen container supply capability. All three of these containers held about 40 gallons of product each. The hydrogen containers are rated to lose less than 2% of their contents per day through vaporization. The liquid was pumped from the hydrogen container by pressurizing the containers with helium, via a siphon to a vaporizing coil array, thence as a gas via a 1¼ inch line to the engine compartment. The helium expulsion pressure effectively established the vapor pressure in the hydrogen vapor lines. The oxygen system is different, due to availability of components which provided the least development effort. The oxygen container is a dewar as are the hydrogen containers, but one modified to deliver vapor by having an integral vaporizing coil just inside and thermally connected to the outer jacket. The vapor is supply by internal pressure regulation at about 80 pounds per square inch. The oxygen pressure was then further regulated by another regulator to about 10 pounds per square inch to a 1¼ inch line and taken to the engine compartment. The gases were then again regulated at the engine compartment to low values for use in the engine. The separate vaporization systems were, of course, compromises to expedite the development since it is highly desired to integrate all the thermodynamic flows as already described.

Being a closed loop, the special engine components will be described beginning at the simplest point, the exhaust ports. The exhaust, consisting of hydrogen and water vapor is passed from the exhaust ports into a special device, the condenser housing. The hot exhaust mixture is passed over the boiler tubes, of flat seamless tubing silver brazed in place, to vaporize the previously condensed exhaust water and to provide the first stage of exhaust heat withdrawal. The exhaust water vapor condensation is completed by passing the gas misture through a radiator like heat exchanger for transferring heat into the cooling water that has just left the normal cooling radiator. Some of the liquid water is collected from the condenser and pumped by several electric pumps to a common point. The drier hydrogen is then fed to a centrifugal separator to spin out remaining water droplets which are also collected and pumped by another electric pump to the same common point mentioned above. This water is then fed to the boiler heated by the hot exhaust gas mixture as described above. The water vapor is then fed out the system exhaust pipe to the side of the truck. No special tail pipe is required since the exhaust is nontoxic, no muffler is needed since the exhaust tested consists mostly of a slight rushing sound with an accompanying sputtering component of still lower strength.

Upon leaving the centrifugal separator, the now relatively dry hydrogen is augmented with new hydrogen from the cryogenic source at about one-quarter pound per square inch. Blowby gases, mostly hydrogen and steam, are also added from the sealed crankcase via the rocker arm housing cover into the separator.

The tangential flow is stopped by passing the gas past a set of vanes. The gas flow is reversed and flows through a restriction to provide a low pressure region at which oxygen is admitted at reduced pressure in a manner akin to fuel carburetion in normal gasoline engine operation. The term "oxyburetor" has been coined to describe this functional device and is etymologically correct in describing the substance and process by similarity to that described by the word "carburetor". The combustible mixture now consisting of mostly hydrogen, oxygen, and some water vapor, is passed to a stock throttle plate where flow is controlled via linkage from the foot control as in normal gasoline operation.

Ignition and valve timing were stock. However, to prevent explosions in the distributor, due to hydrogen leaking from the crankcase into the distributor, it was sealed with silicone sealant and pressurized slightly with pure hydrogen which, in the absence of oxygen, is quite inert and a generally desirable filler gas for both the crankcase and distributor. Although untried, the thermodynamic value of using the incoming hydrogen to cool the crankcase should prove to be favorable. Insufficient operational time has been logged to prove or disprove the possibility of hydrogen embrittlement. The sealing of the crankcase was relatively easy with the exception of the front pulley seal, which had been cut away by a defective casting, and required replacement.

Specific performance data was not quantitatively developed due to a general lack of instrumentation and was not as encouraging as the experimentors had hoped due to experimental apparatus limitations; however, the following observations are recorded. The pickup truck had limited acceleration, top speed and hill climbing ability. It had, as did the Model A, the good qualities of easy starting (when properly set up), quiet operation and the primary nonpolluting characteristics. The pickup truck's performance limitations were probably all the result of an inadequate oxygen delivery capability. The oxygen source was, as previously noted, from a self-vaporizing source (Linde LC-3G) which has a long vapor path with fairly high flow impedance which precluded the required rate, thus limiting the input to a fraction of that required. This performance, although weak, was encouraging in one respect. Since the hydrogen consumption rate, by volume, is twice the oxygen rate, by volume, the performance achieved was apparently economical in terms of hydrogen rate for hydrogen actually consumed (rather than leaked). The hydrogen system impedance is low enough not to be a limiting factor indeed, the real useful hydrogen flow rate was apparently quite low since the actual combustion rate is dependent upon the oxygen rate.

Estimated operational economy in terms of fuel cost per distance generally extrapolates to favorable values for the pickup. Actual oxygen depletion rate was not observable on the float gauge during the tests and, based on observation of frost accumulation on the tank outside over the vaporization coils, probably, was never more than about 200 cubic feet per hour. Since the vehicle realized about 40 miles per hour during this consumption rate, the hydrogen consumption rate was estimated at about 20 miles per pound.

## FUTURE DEVELOPMENT

The liquid hydrogen industry has forecast product cost for large scale transportation fuel and general use liquid hydrogen. These range about 10¢ per pound production cost at the plant, using conventional, present-day sources of energy and product, i.e. natural gas, water and electrical energy using the steam/methane reforming process.[7]

Radically new and undeveloped hydrogen production techniques could include fusion generated electrolysis energy, direct nuclear energized magnetohydrodynamic dissociation of water or similar exotic processes, and these could reduce the cost per pound still more.

Using the specifically measured consumption rate for the Model "A" of 10 miles per pound, the driving cost would be $0.01 per mile. The pickup truck's rate of 20 miles per pound is estimated, but seems more favorable.

The experimentors have given much thought to the many problems which the full application of this fueling system pose. Three major industries face significant disruption were the system to be developed and implemented. The petroleum industry, of course, would face a major problem. Gasoline production would, presumably, decline but the industry would be able to market hydrogen as such or crude and partially refined product as a source of hydrogen to the cryogenic industry. The automotive industry would face a significant technology development in production and maintenance design. The rapidly growing and active cryogenic industry would be required to expand production and develop higher efficiency conversion.

The two most serious factors affecting the development of this technique are fuel volume and fuel logistic development. For a range equal to gasoline operation range, the total volume of the hydrogen storage, including insulation, may have to be up to two times as large, depending on the tank geometry and insulation technology. Some of this volume can be taken from space previously occupied by the exhaust system, but some sacrifice of trunk space or styling may still be necessary. Onboard separation of air into oxygen and its other components may be developed in the final versions of the vehicle.[8] This would increase equipment complexity and weight, but would greatly reduce carried cryogenic weights. A suitable step to develop public acceptance and desire for this concept will be to construct a cosmetically satisfactory conversion to a current production model so that performance is equal to or better than the gasoline configuration. In order to probe the realities of the logistic development, the developers intend to prepare an easily transportable operational mock-up of a future service station equipped to handle the fuel in a safe and efficient manner by a typical service station employee.

The experimentors have also given much thought to the general safety problem. Their conclusion is, basically, that the liquid hydrogen danger is quite definitely different, but not greater than gasoline danger.[9] Techniques have been contemplated for dealing with the hazard raised by the ambient vaporization of the liquid hydrogen. This hazard may be directly minimized by use of the solid phase to take advantage of the latent heat of fusion to stave off vaporization, simple burning consumption, catalytic consumption or even some useful process such as reaction in a small fuel cell for battery charging. The withdrawal of the latent heat of fusion to stave off ambient heat leak vaporization of the liquid hydrogen would allow for a few days shut down without vaporization.

The developers expect to pursue the effort in a series of steps leading to a conclusive basis for establishing the total practicality of the concept of burning only hydrogen and oxygen in a standard automotive otto cycle engine basically designed for air and gasoline combustion.

## REFERENCES


[1]J. M. Seymour, Jr., U.S. Patent #1,275,481, August 13, 1918.

[2]R. A. Erren, British Patent #353,570, July 30, 1931.

[3]R. A. Erren, British Patent #364,179, January 7, 1932.

[4]R. A. Erren, U.S. Patent #2,183,674, December 19, 1939.

[5]R. E. Quigley, Jr., **et al.**, U.S. Patent #3,471,274, October 7, 1969.

[6]E. S. Starkman **et al.**, "Alternate fuels for control of engine emission", February 1970 (Vol. 20, #2), Journal of the Air Pollution Control Association, pages 87 to 92.

[7]J. E. Johnson, "Economics of Large Scale Liquid Hydrogen Production", A Linde division of Union Carbide Co., paper presented at the cryogenic engineering conference in June, 1966 at Boulder, Colorado.

[8]E. M. Kirkpatrick, U.S. Patent #2,960,834, November 22, 1960.

[9]A. D. Little, "Hydrogen handling Handbook".

[10]"1st Annual Report of the Perris Smogless Automobile Association", P.O. Box 892, Perris, CA 92370.

[11]Code of Federal Regulations, USA Title 49, specifically 49 CFR 170.13, 49 CFR 171.8, 49 CFR 177.834, 49 CFR 177.840.


# Hydrogen Experiment

*From Alternative Sources of Energy #10*

**By Bill Vasilakes**

Martin Johanneson has an active mind; he can't talk for a 5-minute period without spouting out a dozen new and sometimes goofy ideas.

He's the guy that bull-dozed a trench in the top of a hill, covered that with oak poles, straw and plaster and dirt, and has a "root-cellar" about 32′ x 100′ in which he has stored bees. They stay at about 40°F in there, even in mid-winter like now, and eat less honey I guess. Anyway, some beemen are pleased with it.

One evening, Martin came to me because I'm a science teacher, and he wanted to know about hydrogen as a fuel,—how we could make it. I told him about electrolysis and his idea was to break down water with a whole series of alternators or generators along a stream bank . . . —stuff salvaged from old cars. He wasn't thinking about pollution but I was, and the beauty of that scheme hit me. Here is an alternative to the storage battery, —just take any old form of energy and turn it into $H_2$. Use the $H_2$ for whatever purposes and get back a dab of $H_2O$ . . . Beautiful!

Then the question came up, would $H_2$ fuel a car? Would it knock, blow up . . . what would it do? We mentally compared it with propane and just guessed it would work. Martin wondered how we could get some hydrogen so I told him how to generate it with acid and zinc. And we agreed we'd do it. I brought a handful of zinc and a half-pint of acid from school and waited for Martin to show up.

One day, he pulled up into the driveway, dived into the back of his station-wagon and came up with a valve, some tubing and tools, and all the time asking if I had the stuff to make $H_2$. When I nodded, he cut his fuel line and stuck in the valve.

He also had a couple of huge balloons and a gallon jug. We threw the zinc in the jug and stretched the mouth of the balloon over the top of the jug. In a few minutes, the balloon filled up and stretched tight to about 18″ in diameter.

Now, we hadn't ever done anything like this before so we didn't know what would happen. We wanted to know exactly when the gasoline would be depleted in the carburetor and when the $H_2$ was going in. So the idea was to start the car on gasoline, let it get running good, then shut off the valve to cut off the gasoline. Thus, it would run so long as there was gasoline in the carburetor, then it would die.

Well, we tried the cut-off valve first, to see what the engine would sound like when it was out of gas. Then we started the car again for the test. Martin started the car while I held that big balloon of $H_2$ over the carburetor; I turned off the gasoline valve and pretty soon, the engine began to die, so I squirted some $H_2$ over the carburetor; and you wouldn't believe it!! That rattledy old engine just turned instantly into the quietest purring thing you ever heard! Lordy it was smooth! I just stood there with my mouth open, pouring that $H_2$ into the carburetor until it was gone and the engine stopped. Martin was still in the car; he heard the difference but couldn't believe the car was really running under power.

So we filled the balloon again and this time Martin did the squirting. And do I mean "SQUIRT"!!! The car started to die down and Martin would squirt in a bunch!! The car would go "vrroooom" and then begin to die down and Martin would give it another squirt so that sweet old engine vroooomed and vroooomed with each squirt until it was out of $H_2$.

I got out of the car and Martin was hopping up and down on his knit cap,—first one foot, then the other—and yelling at the top of his lungs: "Bill! It works! The darn thing works!!" and I almost had to laugh at him, or it, or something.

But lots of questions come to mind: storage, safety, low BTU's, etc., so it isn't likely anyone is going to take off cross-country on $H_2$ for a while. The one thing that doesn't match with what I've read is the way that the engine smoothed down and ran so quietly. I had thought there would be a big hassle with carburetion and all that. Maybe it was running inefficiently, but you'd never know it!

Our next step is to collect some generators and turn them with water-power from some small streams that go by my place. I am doing headwork on some waterscoops that will create enough fall to do the work of turning the generators. I have 3 sons all interested in cars and always changing generators and starters and junking this and that and Martin has a brother with a yardful of wrecks. Anyway, I have not measured the fall anywhere but I figured I'd just cascade the stream with those scoops.

The small drop of the water ought to furnish the power. From each generator, I'd run wires to an electrolytic tank and this is where the hydrogen would be collected in a large balloon. As the balloon fills and swells, it will take up more room and keep getting bigger until it presses against a switch. When that happens, a pump will start and pump until all the $H_2$ is out of the balloon and the weight of the balloon will press an "off" switch.

For home power, I'm in favor of an inverted tank over a water well. Just pump the $H_2$ into the middle of the well and let it bubble up to the inside of the tank. The weight of the tank,—or anything you may pile on top of the well—would furnish the pressure to send the $H_2$ to the house or wherever.

For auto use,—lordy, so many ways to go—no internal combustion engine that I know of was really designed for $H_2$ fuel. So we don't know what can be done with efficiency there. Maybe an external combustion engine is the way to go with $H_2$ heating the boiler. Fuel cells . . . . well, there's lots to look into.

You wanted to know what else we've been doing. Well, we tried some more hydrogen in Martin's slant-six Dodge. You know, we knew what we had heard. That car purred when we squirted $H_2$ into it. So the talk about knocking, about blowing up, about knocking holes in the pistons, about the danger. All of this and a lot of kidding made us wonder if we were really wrong.

So I bought a tank of Hydrogen from Twin City Oxygen Co. in Mpls. It was a red tank about 4 ft. long and weighed about 50 lbs., maybe more. It held about 5 cu. ft. and cost about $6—can't remember exactly.

Had to rent a regulator, too. The $H_2$ is under pressure in the tank—about 2400 psi. Yeah—that's twenty-four hundred lbs. per sq. inch. Just any old regulator won't work.

And we bought some thick-walled hose, clamps, and we already had one of Bates' adapter kits. You hook it into the carburetor and it'll burn methane from chicken shit or other sources.

I'll try to make this sound simple—because it was. We stood the tank in the back seat. My nephew Mac held it upright. We screwed a stub copper pipe onto the regulator. Just think of a copper tube going from a copper tube going from a bottled gas regulator to the stove. Only our copper tube wasn't that long. Instead we clamped the thick-walled rubber hose (hydraulic hose) onto the stub. The other end of the hose went to Bates' adapter. And another hose went from there to the carburetor.

So picture us. Mac hanging onto the top of the tank holding it to keep it from toppling. I was in the back seat with them—the tank between me and Mac. I was holding Bates' adapter and I was to control the valve on the regulator. Martin drove. But first, we ran it at idle—or tried to. I opened the tank. I suppose you know that you don't crack open that valve with a twist. There's 2400 psi in there and you don't let that hit the regulator too fast. I had turned the valve before we hooked it all up—just to get the feel of it. Now I eased it open and the pressure gauge went up to near 2400 before it stopped—about 2370 maybe.

Then I opened the regulator valve to put about 50 lbs. per sq. in. in the tube going to Bates' adapter. The car ran but sounded troubled, even at idle. I played with the regulator—running the pressure up and down—but not much improvement.

So we all got out of the car and raised the hood and listened to the engine be sick. Then Martin did something that amazes me yet. I told you he was a mechanical genius (among other things) didn't I? Well, he took off the air cleaner and started plugging it to shut off the air. But it didn't plug with stuffing rags into the air cleaner cracks. So we took it off and criss-crossed electrician's tape over the carburetor, leaving just a bit of a hole for air.

Now how did Martin know enough to do that? Look, I've got a Master's degree in Chemistry, but it was Martin who saw the trouble. If he hadn't been there, I'd still be scratching my educated head wondering what went wrong.

Sure, I can explain it now, after Martin fixed it. It's simple—it takes two hydrogen atoms to react with one oxygen atom when the two burn. So you have to put more hydrogen than oxygen into the carburetor. We are used to lots of oxygen and a tiny squirt of fuel going in there. So the ordinary carburetor is built all wrong for hydrogen. Unless you choke it off as Martin did. Even then, there ought to be a better way.

Air, of course is only about 20% oxygen so it ends up with about 5 parts of air going into the carburetor with 2 parts of hydrogen. The four parts of nitrogen in the air just go along for the ride. Don't I sound learned? So I scratch my head while Martin figures it out.

Back in the car and off we go heading south for a long climb up the hill. The pressure was dropped to about 2200 in the tank. We must have squirted hydrogen all over the place until the carburetor was fixed (choked). But no danger. At least our voices weren't high-pitched so there couldn't have been that much hydrogen around.

What do I mean? Well, if you breathe hydrogen into your lungs, then talk, your voice goes up in pitch. You sound like a girl when you talk. The lightness of the hydrogen gas causes faster vibrations or something in the vocal cords. Come to think of it . . . . maybe there was a lot of hydrogen. Maybe hydrogen can float into your ears and make high sounds seem normal. Maybe in a hydrogen gas atmosphere, a girl's voice would sound bass.

Yeah—back to the car. Hydrogen isn't that dangerous. No worse, so far as I can see, than any other fuel. To explode, it has to have oxygen. But in the tank what's to worry. Except all that pressure.

On the level ground, the car ran well, but the pick-up—the acceleration was sluggish. When we started, I had the regulator on 50 psi. With the sluggishness, I ran it slowly up to 100 psi. At that pressure, the rubber hose was as stiff as a steel rod.

And Bates' adapter started singing

songs. It buzzed and chattered and clattered so loud you couldn't hear. We had to yell at each other. We didn't like that noise because we surely wanted to hear those pings in the engine that so many people predicted we would hear with the engine under a load.

So we stopped the car and redid everything so that the hose ran under the hood to Bates' adapter, which we wired to a brace above the engine.

We started up again—and blessed quiet with the buzzer under the hood. But we couldn't hear it. Didn't hear much except the tires on gravel. Even with the windows up—or down—we did not even hear that engine ping. It ran quieter than usual—and that's all.

And Martin tried to make it ping. He stopped on the hill, put it in 2nd gear and make it lug. No ping. It didn't even buck. It just smoothly gained acceleration from a dead stop and in 2nd gear. Cars don't do that on gasoline.

Martin tried various things as I sat watching the pressure dwindle in the tank. He gunned it in low, lugged it in high—tried starting and stopping, braking it down with the clutch engaged. I guarantee you, gentlemen, a slant-six Dodge engine will not ping when it's running on hydrogen.

We went down the road about 5 miles before we ran out of hydrogen. With the amount we wasted, I suppose we can figure about $1 per mile for that kind of fuel. And with that silly carburetion.

Martin and I were somewhat depressed when the trip ended. No matter what improvements I could envision on carburetion, there's no way to consider $H_2$ as a practical fuel for mobile units. Oh sure, you can go a few miles with $H_2$ tanks. But the weight of the tanks and the amount you can squeeze into them just isn't worthwhile. Even if you generate $H_2$ gas and get it free from the wind, you'd be limited in the range your vehicle could go. It's not the cost, it's the range.

I'd like to hop in and go to Minneapolis from home—that's 280 miles. Not with hydrogen. I'll bet the range won't be much better than with batteries in an electric car.

But for short range—to work and back—a few miles. Why not hydrogen?

Unless Martin thinks of a way out of this dead-end.

Nice talking to you, John
Sincerely, Bill

*– Bill Vasilakes*
*Lengby, Minn. 56651*

## HEAT PUMP

Wouldn't it be nice to have a machine that could extract heat from the cold winter air and bring it indoors, warming the inside at the expense of the outdoors? Such a device exists. It's called a heat pump, and it doesn't involve a violation of the law of thermodynamics, though the idea may seem magical. A heat pump can be pictured as an air conditioner working in reverse. In fact, both air conditioners and refrigerators are examples of heat pumps.

The working principle of all three is the same. In each, a liquified gas is allowed to expand and vaporize. (An expanding gas absorbs heat, which is why spray from an aerosol can feels cold.) The cold gas is circulated through a set of coils where it absorbs heat from the warmer surroundings. Next, the warm gas is compressed. As its pressure rises, the gas gets hotter. (Anyone who has used a bicycle pump can understand this.) The hot gas then gives up its heat as it passes through a second set of coils. When it has lost enough heat, the pressure turns the gas back into a liquid, and the cycle starts all over again. Heat has been transferred from a cool place (the first set of coils) to a warm place (the second set of coils), hence the name "heat pump". Physics students will recognize the process as a heat engine cycle run in reverse. The idea was first proposed by the English physicist, Lord Kelvin, in 1852.

The price paid to make heat flow "backwards" is the energy required to circulate the gas and run the compressor. Typical heat pumps can deliver from three to four kilowatt-hours worth of heat for every kilowatt-hour used. The ratio is called the "coefficient of performance" or "COP" for short. Obviously, electricity that is used to run a heat pump gives more heat than the same amount of electricity would in a electric heater. Yet heat pumps have not been used much. The main reason is that a heat pump installation is more complex than an ordinary heating system, so it's first cost is greater. If fuel costs are low, it takes a heat pump a long time to show an advantage over ordinary heating systems. Now that the price of energy is rising, people are becoming more interested in heat pumps. Most of the companies that make refrigerating and air conditioning equipment also offer heat pumps.

The heat pump has many interesting applications. In the summertime, a heat pump can be reversed to act as an air conditioner. And the source of heat that a heat pump uses does not have to be the outside air. It can be a nearby stream or lake, or the cooling water from a blast furnace, or almost any other source of low grade heat. Heat pumps have been used to cool a cold storage room, using the heat obtained to warm the rest of the building.

A heat pump could be very useful in combination with a solar heating system. It would take the low temperature heat delivered by the solar collector, concentrate it, and make it available at much higher temperatures. In this way a house could be warmed to comfortable temperatures in the dead of winter by solar collectors that were only a few degrees above the temperature of the outside air. Adding a swimming pool to the system makes other possibilities available. A pool full of water makes an excellent heat reservoir. In the winter the pool, warmed by solar collectors, could serve as a heat source for the heat pump. In the summer, the heat pump could warm the pool while cooling the house. Such a system would require the use of some outside electric power but it would make more efficient use of the heat gathered than an all solar system.

# An Experimental Heat Pump System

By Richard La Rosa

The original idea was to gain building and operating experience by heating an extension of a private house by means of solar energy. The extension consists of a bedroom and an insulated, unfinished bathroom built over a full cellar. Both rooms have three exterior walls and it takes a steady kilowatt from an electric heater to keep them warm in 10°F weather.

The two most important problems considered at the outset were:

1. Availability of an inexpensive heat reservoir massive enough to carry the system through a week with no sunshine.
2. Accessibility of solar heat collectors for construction and service.

## THE COLLECTORS

It was decided that three 4 ft. by 8 ft. collector panels could be mounted on a 7 ft. high scaffold erected behind the swimming pool. The collector consists of aluminum sheets with closed convolutions on three-inch centers. (See A, B, & C.) The exposed metal surface is painted flat black. The sheets are enclosed in plywood boxes with insulated backs and .004 inch thick cast vinyl windows in front. Water is pumped to plastic feed pipes which run along the tops of the boxes. Water is then fed to each corrugation individually by means of small plastic tubes glued to the main feed pipe. The collector plates are inclined 25° from vertical and face almost due south. The water runs down the closed convolutions (which are almost vertical) and is returned to the pool by troughs and pipes.

The present collector pump is belt-driven by a washing machine motor and consumes 600 watts. This power is fa r in excess of what is required to lift a few gallons per minute 7 feet above pool water level.

## FACTS AND FIGURES

When sunlight strikes the collector at normal incidence, the rate of heat collection (i.e., the power flowing into the heat collector) should be 1 kilowatt per square meter, or .092 kw per square foot. The 96 square feet should therefore absorb power at a rate of 8.8 kw. A power flow of 1 kw is equivalent to a heat energy collection rate of 3400 BTU per hour, so this collector should have a collection rate of 30,000 BTU per hour at noon on a sunny day.

The heat storage reservoir (the swimming pool) contains 62,000 lb of water. One BTU raises the temperature of one lb of water one degree Fahrenheit. On a sunny winter day, the pool water temperature increases 1°F, indicating that 62,000 BTU are collected.

On that same sunny winter day, the collector pump runs for 7 hours. The average rate of heat collection of 62,000/7 = 8800 BTU per hour. The average collection rate during the day is 29% of the peak rate expected in early morning and late afternoon, and also to the increased reflection from the vinyl windows as the rays depart from perpendicular incidence.

The cast vinyl windows collect dirt and oxide film after about one season and this cannot be scrubbed off. Calendered (rolled) vinyl is said to be superior in this respect. Glass would be far superior, but its weight and ease of breakage preclude its use. Someone has suggested "Tedlar" but I have to check this out.

The collector pump is presently controlled by a thermostat in a heat collector box which senses when the radiant heat collection rate exceeds the losses to the outside air.

## STORAGE

The swimming pool is the usual above-ground type with a steel frame, steel wall, and plastic liner. It is 18 ft. diameter by 4 ft high, holds 1000 cu ft of water. Each 1°F represents 62,000 BTU, or about one day of heating for the extension in cold weather.

Preliminary calculations of ground conductivity losses showed that there was little possibility of maintaining a water temperature high enough to heat the house directly. We decided to operate with the reservoir temperature at any value above freezing, even though a heat pump is required to transfer heat against a temperature differential.

The pool is located 20 ft. behind the house. Heavy wall polyethylene pipe, one inch diameter, was laid in a two ft. deep ditch. The two pipes are insulated from each other and from the ground by means of blocks of polyethylene foam and styrofoam chips stuffed into plastic bottles. Sheets of plastic cut from jugs were laid over the insulation and the trench was back-filled with earth.

1 Solar Heat Collector
2 Insulated swimming pool

At the pool end, the pipes are brought up close to the wall and pass over the top. One pipe is connected to a pump and the other pipe is connected to a length of hose which leads the discharge away from the pump inlet. At the house end, the pipes curve up through a cellar window and pass through the floor of the extension. The return line is connected to the bottom of a washing machine tub and the inlet line passes over the top of the tub and is connected to the water passages of the heat pump.

## THE HEAT PUMP

The heat pump is a discarded air conditioner. Its evaporator coil is bent away from the unit and submerged in water. The incoming pool water is forced to flow between the fins of the evaporator, following the path formerly taken by air-conditioned room air. The evaporator is located in a sawed-off plastic gabage pail, which is placed inside the washing machine tub. The water fills the pail overflows into the tub, and returns by gravity to the pool. The bottom of the tub is about two ft above pool water level.

The transfer pump in the pool requires 100 watts and pumps less than 1 gallon per minute through the heat exchanger. It was found that in cold weather, ice formed on the outside of the evaporator coil. This blocked the water passages and prevented heat extraction. A stirring pump outlet was directed toward the evaporator coil. This prevents icing but uses another 100 watts of electrical power.

The refrigeration compressor uses 690 watts and the fan which blows room air over the condenser uses 70 watts. The heat pump therefore uses a total of .96 kw of electrical power, of which .86 kw passes directly into the room. By measuring the rate of temperature drop of the water, we have determined that 1.38 kw of heat is extracted from the water. The total heat into the room is .86 > 1.38 = 2.24 kw. The coefficient of performance is 2.24/.96 = 2.3. Only .86 kw goes into the room. The transfer pump's loss heats the pool insulation and is probably not recovered.

From a household economic standpoint the system is definitely superior to electric heat and probably superior to gas or oil. From an ecological standpoint, the heat pump barely makes up for the poor efficiency with which electricity is derived from fossil fuels.

## INSULATION

The top of the pool was insulated by sheet of "bubble plastic"—two layers of plastic bonded together with air pockets between the layers, used for packing and shipping equipment. This bubble plastic is laid on top of a nylon safety net. The weight of the bubble plastic is supported by a 15 ft by 4 ft air pillow surrounded by plastic jugs and pieces of styrofoam and

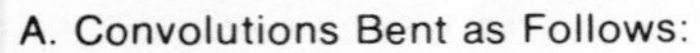

A. Convolutions Bent as Follows:

*Hit wood piece with rubber mallet to force 4 ft wide sheet between 2 x 4's nailed to base*

*aluminum sheet*

*plywood base*

polyethylene foam floating in the pool. Another nylon net is stretched over the bubble plastic and a sheet of black polyethylene is laid on top of the net. Nylon monofilament (40 lbs strength) is passed over the top and held in place by weights.

Plastic bags filled with styrofoam pellets are packed around the side of the pool. The pellets are packing material discarded by an electronics plant and a local motorcycle dealer. The plastic bags must be protected from puncture. This is done by tying 4 ft high pieces of styrofoam around the outside. The pieces are cut from discarded motorcyle shipping cartons. They are full of holes and recessed and by themselves are not good as insulation.

The ground directly below the pool is part of the heat storage. When the pool water drops below 45°F, the heat deep in the earth is actually flowing into the pool via the ground path. The sidewall insulation is made as thick as possible in order to minimize heat loss to the cold ground surface.

## OPERATION

The insulation was put on the pool in late fall of 1972, and the heat collector ran automatically on sunny days. The water temperature dropped steadily and reached a low of 42°F on Jan 1, 1973. It rose to 49°F on Jan 23 and then dropped back to 42°F on Feb 12. From this point on, it rose steadily to 65°F on April 24. The collector motor was unplugged at this time.

The two low temperatures mentioned above followed periods of low sunshine. The peak of 49°F followed a period of steady sunshine. The dependence of water temperature on sunshine indicates that the primary source of heat is the directly collected solar energy rather than the heat stored in the ground from the previous summer.

Measurement of the temperature profiles in the ground would be useful to establish the rate at which solar heat is

1 Heat pump compressor
2 Heat pump condenser & fan
3 Water from pool
4 Heat pump evaporator coil inside plastic garbage pail in washing machine tub

Note: Stirring pump not shown

being wasted into the surrounding ground. We know from some crude calculations that the most severe temperature gradient is at the ground surface beneath the sidewall insulation. During the summer of 1973, we are digging a ditch six inches wide and 18 inches deep, and filling it with styrofoam. A deeper ditch would be better, but clay and tree roots force a compromise at this point. The ditch is about 3 ft from the base of the pool. The wall insulation will be extended to cover the ground surface inside the ditch.

B. Conduit, ½" dia., insert temporarily
Pinch open channel around conduit with smoothed jaws of nail puller

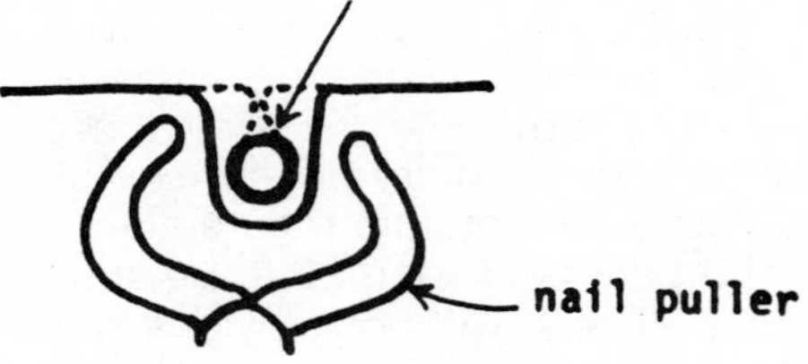


C. CROSS-SECTION OF COLLECTOR LOOKING DOWN FROM THE TOP:

1/4" plywood
BACK
fiberglass
1/4" plywood
aluminum
.004" cast vinyl
FRONT
(Styrofoam would be better -- will not waterlog if box leaks)
3" approx.

Some other improvements which are contemplated for the coming winter are:

1. Increased separation of intake and exhaust lines for the heat collector and transfer pump. Presently, water discharged from the heat pump enters the pool only 3 ft from the intake. It should have been routed to the opposite side of the pool to avoid recirculating cold water back to the house.
2. Smaller motor and better pump for the heat collector.
3. Increased collector area.
4. Large styrofoam floats on top of the water to block convection losses to the top covering.

Readers who wish to comment can contact me at this address:

Richard La Rosa
317 Oak Street
S. Hempstead, NY
11550
Phone: 516-486-7827

*From Alternative Sources of Energy #11*

# SOLAR ENERGY — SOME BASICS

Each day the sunlit side of the Earth receives the energy equivalent of 4,000 trillion kilowatt-hours. By contrast, the whole world uses roughly 80 trillion kilowatt-hours per year. Solar energy is freely available to all. It cannot be cut off by blockade or embargo. Nor does it involve the burning of any scarce fuels or the generation of any pollution. It cannot be exhausted for it will last as long as the sun shines.

With this sort of potential it is no wonder that people are excited by solar energy. However there are disadvantages too. While the total amount of solar energy is great, it is spread over a vast area. To be useful, it must be collected and concentrated. This means that lots of space is needed for mirrors or collectors. Solar energy is also subject to interruption. It is not available on cloudy days or at night. To insure a dependable energy supply, the captured energy must be stored in some way or supplemented with a back-up system. In addition, solar energy is not as flexible in its application as are more conventional sources of power. A fuel tank or an electric motor can be placed almost anywhere, but to use solar energy you need plenty of room in a sunny place in which to spread solar collectors. As long as fuel is cheap, there is no particular reason to suffer these disadvantages for the sake of using solar energy. Now that the cost of energy is rising sharply, solar energy is getting a lot more attention.

The availability of solar energy varies from place to place according to the prevailing weather conditions. In the Southwest, with its high percentage of clear days and nights, solar energy is at its best. In areas like the Pacific Northwest, or northern New York state, it is limited because of frequent clouds and overcast. By consulting an almanac or atlas, or Weather Bureau records, you can find out how much sun you can expect in your area each season.

Solar radiation is measured in **langleys**, a unit named after Samuel Langley who invented instruments for measuring solar radiation. One langley is radiation energy equivalent to one calorie falling on an area of one square centimeter. The measurements are taken on a horizontal surface at ground level. The solar radiation flux varies from zero to 1.5 langleys per minute. One langley per minute is a typical value to expect on a clear day. Since the sun's radiation is measured on a horizontal surface, the experimenter can expect a bit more when he tilts his collector toward the sun so that it is perpendicular to the sun's rays.

A solar collector is a device intended to absorb heat from the sun. Any object left lying in the sun will do this to some extent, but a solar collector is designed to collect and transfer as much heat as possible with minimum losses. The heart of a collector consists of coils or channels containing a fluid (usually air or water) so that the absorbed heat may be easily transferred to where it will be used. By focusing the sun's rays with lenses or mirrors, much higher temperatures can be attained. Such a focusing type collector must follow the sun closely. It is most effective on clear days. On a hazy day the sun's rays are scattered and the focusing collector, which "sees" only a small part of the sky, can pick up only a small percentage of them.

The alternative to the focusing collector is the flat plate collector. The fluid-filled coils or channels are spread flat and enclosed in a shallow box. The box is insulated and covered with glass to cut heat losses. It is best at supplying large amounts of heat at low temperatures. While the flat plate collector does not reach such high temperatures as the focusing collector, it does not need a sun tracking mechanism. Since it absorbs radiation from the entire sky, it works well even on hazy days. In addition, it is more rugged and easier to construct than the focusing collector. For these reasons, the flat plate collector is the one most often used, except when very high temperatures are needed.

A focusing collector is aimed directly at the sun. On the other hand, a flat plate collector is usually fixed in position. Since the height of the sun's path across the southern sky varies with the season, the tilt of the collector must be a compromise. The sun is highest in summer, lowest in winter, and midway between during spring and fall. The designer can take advantage of this by tilting the collector so that it faces the sun squarely during the time of year when energy is most needed. For instance, a collector in a solar home heating system will be tilted to face the low winter sun so that the collector will work best during the coldest part of the year. The coils of a collector are nearly always blackened to make them more efficient at absorbing light and converting it to heat. Unfortunately, black is also good at radiating heat, and most of this re-radiated heat is lost. To stop this, some collectors use selective surfaces. These are coatings which are good at absorbing light, but poor at radiating heat. The use of selective surfaces makes a collector much more efficient. Unfortunately the making of selective surfaces is beyond the amateur at present.

The amount of energy gathered by a collector depends to some extent on the flow rate of that heat transfer fluid through it. If the rate of flow is slow, the fluid has plenty of time to reach a fairly high temperature. On the other hand, the hotter the collector is allowed to get, the more heat it will lose to its surroundings. When the rate of flow is fast, the fluid does not warm up very much. However, a larger amount of fluid is warmed, and since the losses are less in this mode, the total amount of energy delivered by the collector is greater.

Most applications of solar energy require some means of storing the heat gathered. This can be as simple as a large tank of water. Other methods that have been used include bins of stones and stacks of water-filled jugs around which warm air is allowed to circulate. These methods all make use of the capacity of a large bulk of material to absorb and retain heat. Another method depends on heat of fusion. Materials are capable of absorbing large amounts of heat as they melt and then releasing it as they resolidify. Sodium sulfate, or Glauber's salt, is a substance that has been used for this purpose. It melts at 32°C (77°F). By using other salts, and by combining them in various proportions, it is possible to get mixtures with lower melting points. Such mixtures are called eutectic. Systems using molten salts have the advantage of compactness, though they are more expensive than the other methods mentioned.

In the long term, solar energy will almost certainly be one of our major sources of energy. In the shon term, solar energy can be used to stretch our supplies of fossil fuel. Right now we burn large amounts of our fossil fuels to provide low temperature heat, as in home heating. This could be done just as well by solar energy, which would free large amounts of fuel for tasks requiring concentrated high temperature heat.

# Solar Distillation

By Dr. E. A. Farber
Director, Solar Energy &
Energy Conversion Laboratory
University of Florida
Gainesville, Florida 32611

Fresh water is a problem in many areas of the world, not because water is not available, but because the solid content of the available water is too high to make it drinkable or useable for irrigation.

One method of producing fresh water from either salt water, contaminated water, or even liquid wastes from houses and communities is by means of a solar still.

A solar still has a basin which contains the salt water or contaminated water or waste water. This basin is covered by inclined sheet of glass with a collecting trough at its bottom. The sun shining through this glass will heat the water and vaporize some of it, then the steam coming in contact with the glass will condense on the glass and run down and collect as distilled water in the trough where it can be led to a collecting container. The glass cover can be either sloping south only or it can be a roof type construction.

The stills are usually oriented along the east west line with the glass inclined with the horizontal at 30 to 45 degrees. Depending on whether the water is shallow or three to four feet deep classifies these stills as shallow basin or deep basin designs.

Plastics can be used instead of the glass but usually will form drop wise condensation which reflects solar energy on the glass and, therefore, the yield from these stills is not as satisfactory. One can expect from ½ to one pound of fresh water from each square foot of pan area and this is the equivalent of up to two inches of rainfall per week.

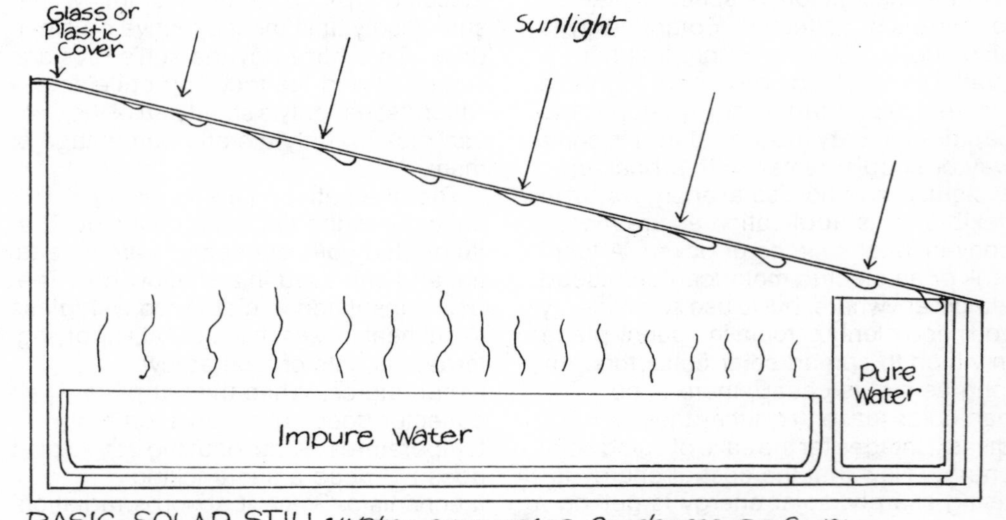


BASIC SOLAR STILL; water evaporates, Condenses on Cover and runs into trough.

## EARTH STILL

Here is a very simple solar still that anyone can make. What's more, it could save your life someday. You need only a sheet of clear plastic film, a can or cup, and a shovel. Dig a hole in the ground about a foot deep and about three feet across. Place the can in the middle of the hole and cover the hole with the plastic. Place rocks around the edges to hold the plastic in place and then put a small rock on the plastic so that it forms a depression right over the can. The soil holds water like a sponge, even in dry areas. The heat of the sun drives the water out of the soil and turns it into vapor. The vapor condenses on the inside of the plastic film, runs down the slope, and drips into the can. The best part about this still is that it is so portable.

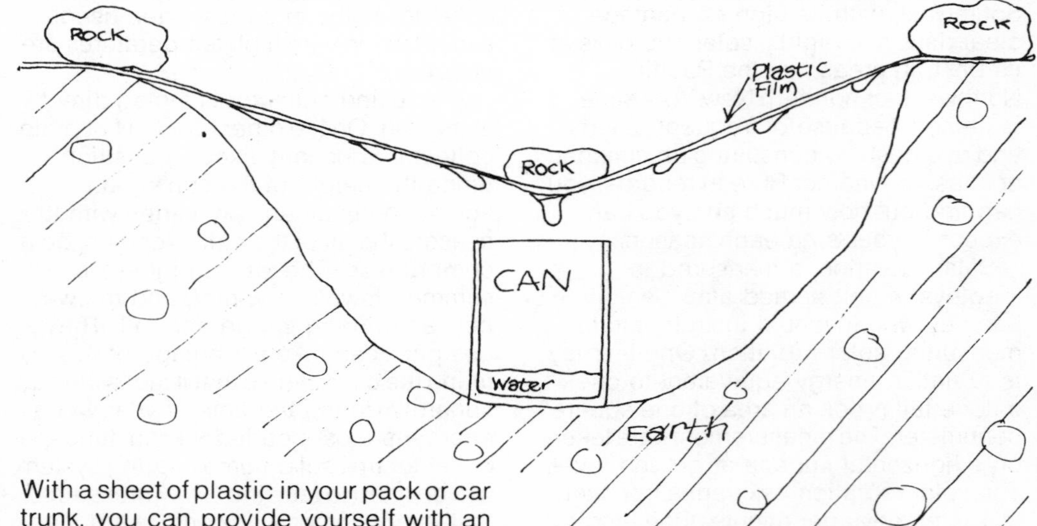


With a sheet of plastic in your pack or car trunk, you can provide yourself with an emergency supply of drinkable water almost anywhere.

# Solar Water Heating

By Dr. E. A. Farber
Director, Solar Energy &
Energy Conversion Laboratory
University of Florida
Gainesville, Florida 32611

Water can be heated by solar energy by many different methods, varying from the very simple to the very sophisticated, expensive, but more efficient.

One of the simplest methods is to take a tank and set it in the sun, or a pipe, or a garden hose spread out on the ground with the sun shining on it. It will provide very hot water in a relatively short time.

The more common method used in the United States is a flat plate collector which consists of a box which can be either wood or metal with a metal sheet inside to which tubes are soldered in sinusoidal arrangements spaced about four to six inches apart. The tubes are usually from 4″ to 6″ apart. The tubes are usually from ¾″ to 1″ in diameter and there must be thermal contact between the plate and the tube. This is important so that the heat which is absorbed by the plate can easily flow to the tube and through the tube into the water. The tube-plate arrangement is painted with a good absorbing paint, usually commercial flat blacks are satisfactory.

The tube-plate arrangement is supported by point supports to reduce the heat losses and has either about one inch air space between it and the back of the box or has one or two inches of insulation in the box.

The box is covered by one or more layers of glass, depending on whether it is used in southern or more northern climates. It is important to reduce the heat losses further north. The glass should be one with low iron content which can be determined by looking at the cutting edge which should be colorless or light bluish and not green which indicates high iron content. Plastic sheets can be used instead of glass but do not have the trapping properties of glass meaning that the short wave radiation from the sun penetrates very readily while the long wave radiation given off by the hot surfaces does not. Also, plastics usually do not last as long due to the ultraviolet effect, the elevated temperature and the wind flexing.

In climates where freezing is not a problem, this collector can be coupled to the tank which contains the domestic hot water. If the tank is two feet or more above the collector, free circulation will take place and no pump is required. In many instances with the collector on the roof, the water tank protrudes through the roof and is camouflaged as a chimney. If it is, however, desirable to have the tank lower than the absorber, a small very inexpensive circulating pump is required. In freezing climates draining or a dual circulation system is required which means that the abosrber is coupled to an outer tank which can have a lid and is insulated. The service hot water tank is submerged inside the outer tank. The right amount of anti-freeze is added to the water in the outer tank so as to prevent freezing for the particular, local, climatic conditions.

The usual dimensions for a single family dwelling is four feet by twelve feet for the collector, and about a one hundred gallon service water tank. It is advisable to install a small electric booster, somewhat like a four thousand watt coil, thermostatically controlled so whenever the sun is unable to provide all the hot water needed in the house, a few times during the year, this booster can provide the difference.

Thousands of these solar water heaters have been operating very satisfactorily in the United States and millions of them around the world, and if properly designed, they will give very satisfactory results. Some studies made by our laboratory indicated that the break even point in the cost of the hot water is about two years. From then on the hot water provided by solar energy is free while other systems will have to be supplied with fuel.

The solar water heater or flat plate collector should be oriented facing south and inclined with a horizontal at an angle equal to the latitude plus 10 degrees. In this manner the collector will be more favorably oriented in the winter when the days are shorter and less favorably in the summer when the days are longer. In this manner it will collect about the same amount of energy all year around.

Some solar heater installations have been in operation in the United States for over forty years.

---

# SOLAR WATER HEATERS

From the Tucson Community Development Design Center — Thanks to Peter Hiatt

Solar water heaters use the sun's radiation to heat water for domestic use. There are generally two main parts to a solar water heating system. The first part is the collector where water is heated by a darkened surface exposed to the sun; and the second is a storage tank where water heated by the sun is held until used. The storage tank is necessarily larger than the normal size for a gas or electric heater since it must hold enough water for use during the night and cloudy times when the sun isn't shining.

There seems to be a difference of opinion on how much hot water is used by people in a day. One source stated that the designer should plan for 20 gallons of hot water per day per person in a middle or upper class household. This is likely to be somewhat more than is necessary for individuals in a lower income household since they are less likely to have such large hot water users such as dishwashers and clotheswashers. In literature received from an Australian company it was indicated that their planning figure was

10 gallons of hot water per day per person; part of the difference in figures may be due to cultural differences.

Many solar hot water systems have supplemental heaters to supply heat to the water in the storage tank when the sun does not supply enough heat due to inclement weather. The Beasely company of Australia estimates that only 30% of the hot water in their country would have to be supplied by gas or electricity and the rest could be produced by the sun; this is an average figure for all of Australia. The percentage extra energy needed to heat water might be smaller in Tucson.

Usually supplemental heaters are controlled automatically by a thermostat and use an electric element to heat the water; one which uses gas to heat would likely be much cheaper to operate. Supplemental heaters would probably not be necessary in Tucson if the hot water system were large enough, i.e., if the storage tank had enough capacity to last through one or two cloudy days in the winter or during the rainy season. However, such a system would be so large and heavy that it would not be economically practical. One could also do without the supplemental heating system if one were willing to change one's life style a little, in other words, not bath or use much hot water on days when the sun is not shining. This type of system would be most inconvenient to household with small children or sick persons and therefore is not practical either. For these reasons, some type of supplemental heater will probably have to be supplied with any solar water heater.

There are two main types of solar hot water heating systems; they are the 'closed box' type and the 'recirculating' type. The **closed box** type of solar water heater is very simple. It consists of a metal box insulated on the sides and bottom which contains several large diameter black pipes. The top of the box is covered by a transparent sheet of plastic. The tops of the large pipes are exposed to the sun and serve as the heat collectors; the hot water is also stored in the large pipes. Hot water is drawn off the top ends of the pipes via a manifold and replacement water is introduced at the bottom ends. The **recirculating** type of solar water heater consists of a flat plate collector and a separate insulated storage tank. Sunlight shines through the glass sheet covering the flat plate collector and strikes a piece of copper painted flat black which has water tubes embedded in the surface. The sunlight is converted into heat which cannot escape by reradiating through the glass since glass is opaque to infra-red radiation. The heat of the plate is carried off by water flowing through the tubes. Since the flat plate collector is located lower than the storage tank, a convection

## Recirculating Type

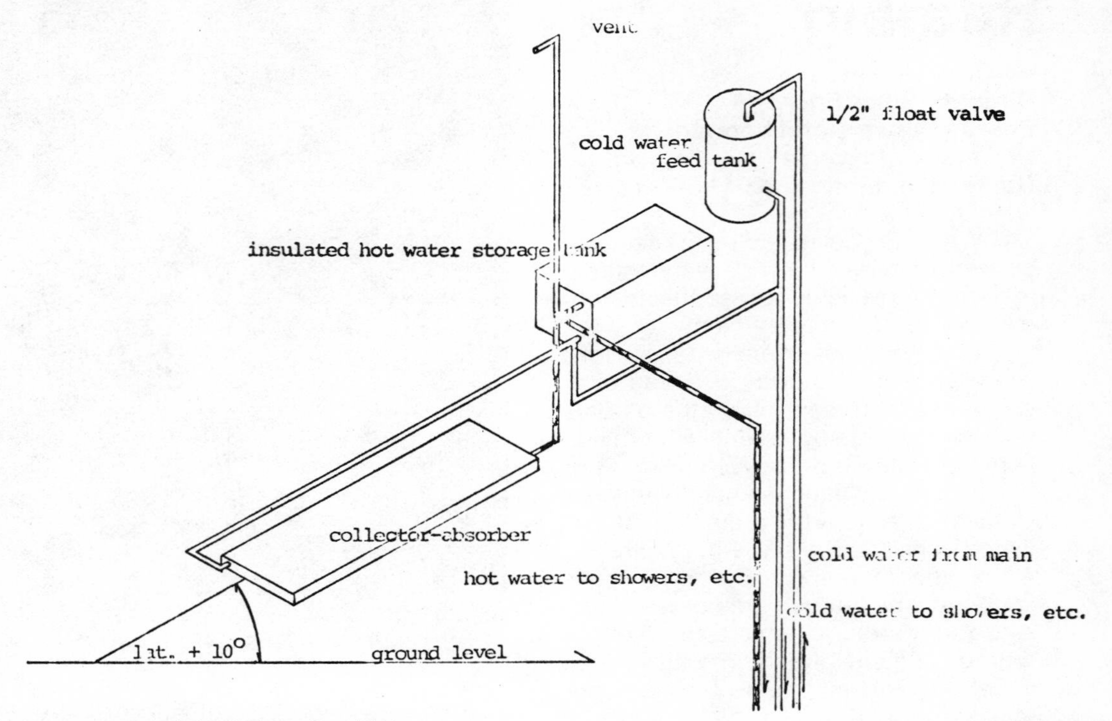


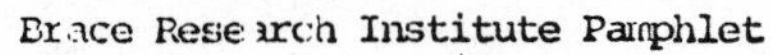
Brace Research Institute Pamphlet

## Closed Box Type

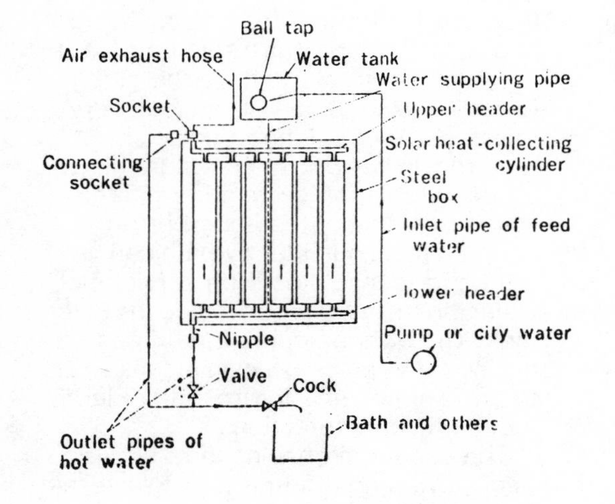


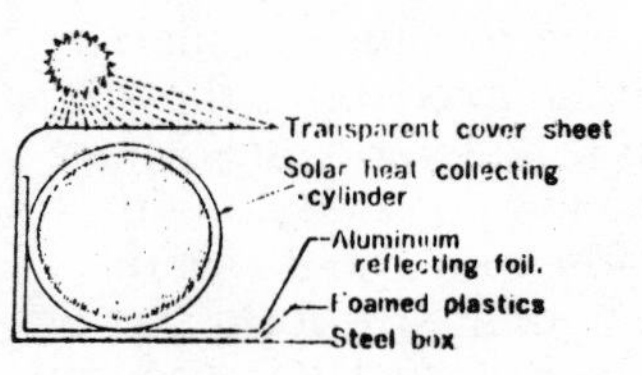

current is set up automatically as hot water is lighter than cold water; the warm water in the collector displaces the cooler water in the storage tank. At night the convection current ceases since the water in the solar collector becomes cooler than the water in the tank; this prevents the loss of heat to the nighttime air.

There are various advantages and disadvantages to both types of systems. The closed box system may lose a large amount of heat at night through the uninsulated plastic cover. Professor John Yellott of ASU mentioned at a solar energy conference in Tempe that a Japanese solar heater of this type was installed on a Phoenix house and proved to be unsatisfactory since it didn't produce enough hot water. The advantage of this type of heater is that the danger of freezing is minimized due to the large amount of heat stored in the water in the tubes. However, there would be a danger of the supply pipes freezing if they were not wrapped in insulating tape. The recirculating type of solar heater while it is not prone to losing its heat at night is more susceptible to freezing in its flat plate collector. One way to prevent this is by 'double glazing' or placing two panes of glass over the black copper plate in the flat plate collector. The second pane of glass provides extra insulation at night and prevents freezing. Some are of the opinion that double glazing is not necessary in a climate such as Tucson's where the air temperature is below freezing for only a few hours at a time.

The placement of the solar collector in a hot water heater is important because it affects the efficiency of the heater. The optimum placement for a solar collector is a position facing due south at an angle of 15 + ° of latitude)° from the horizontal. In Tucson with a latitude of 32° 10′ from the horizontal.

Visually, there is a problem with both these types of solar water heaters since in Tucson many homes have flat roofs. Since the logical placement for them is on the roof, there will be an addition to the silhouette of the house which may or may not be as ugly as the present air cooling units and television aerials. This effect can be minimized by careful placement of the water heater units away from the roof edges and by careful design and placement of parapet walls which could shield the units from view. In addition, the flat plate collectors of the recirculating type of heater could be used as an awning over a window in a southern wall and the storage tank on the roof could be disguised as a false chimney. In homes where the roof is gabled, the storage tank could be located in the attic space and the flat plate collectors placed on the already sloping roof (if it faces south.)

# A Simple Solar Hot Water Heater

By E. W. Short

Nothing new here of course—but it illustrates what can be done just by trying. Many refinements could be added but in the year that my family and I stayed at our little self-constructed home on the beach at Andros Island in the Bahamas our goal was to make do with what was available and simple. Next time we go for any extended time I plan on setting up a used windmill from the states which should be very handy for many many things—water, grinding grain, etc.

This was used in the Bahamas quite successfully. Simple—cheap—no welding used. Clamps were not usually necessary on the plastic pipe and connectors—a friction fit was usually sufficient. The sun will affect the pipe, especially the connectors. The light ones (grey) go first. The black ones last longer. Maybe you could paint them with a heavy black paint to help them last longer.

The water cooled off rapidly with no sun as it would run backwards! Black radiates as well as absorbs. One could use a non return valve, or insulate the storage tank and cover the collector, etc., but it worked well this way. I really don't have any idea of the amount of hot water produced—we always had plenty for showers.

E. W. Short
10500 S. Michigan Ave.
Chicago, Illinois 60628

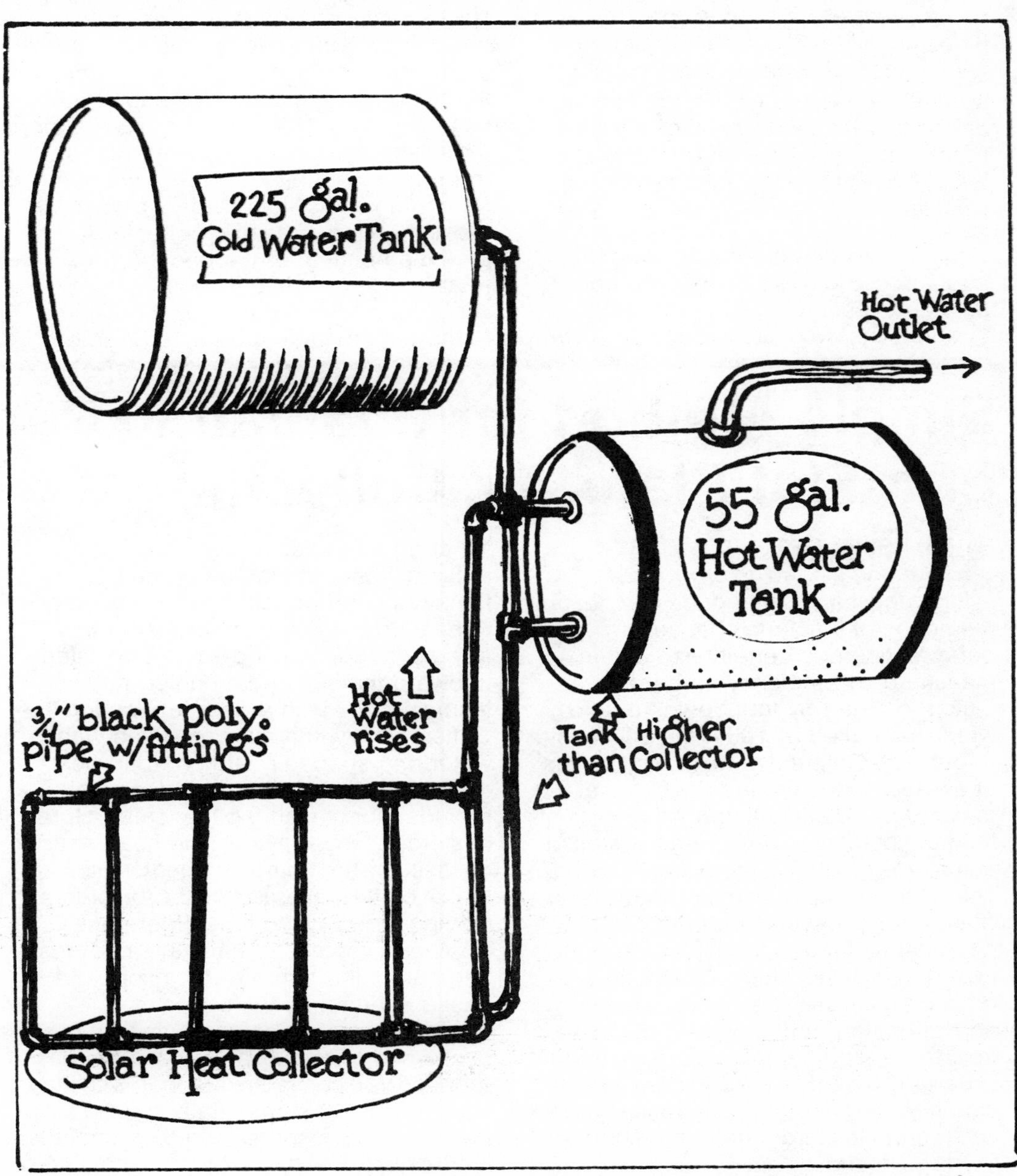

# Swimming Pool Heating

By Dr. E. A. Farber
Director, Solar Energy &
Energy Conversion Laboratory
University of Florida
Gainesville, Florida 32611

Heating of swimming pools is an expensive proposition since swimming pools contain tremendous amounts of water, and thus, require very large amounts of heat to do the job. Swimming pools can be heated by means of solar energy over a period of years cheaper than by any other means; however, the initial equipment is still expensive except for one of the very simple methods.

If a swimming pool, and it can be an olympic size swimming pool, is in the sun, then a relatively inexpensive method is to take a plastic transparent sheet and float it on the surface of the pool so that no air bubbles are trapped below the plastic. If air bubbles are under the plastic, droplets will collect on the plastic and act as little reflectors preventing the solar energy from penetrating into the water. By this relatively simple method, the average pool temperature will be increased by 10°F.

Other methods of heating swimming pools become more expensive. Large surfaces, roofs, etc., can be used to circulate the water and let it trickle down the surface which is heated by the sun. The warming water drains back into the pool. Concrete surfaces or slabs can be used for this purpose or metal sheets exposed to the sun which can be the fence around the pool to provide both solar heating for the pool and privacy for the swimmers. Fences in many areas are required by ordinance to be put around swimming pools, thus, the fence which at the same time is a solar collector will give the heating for the pool at a relatively low additional cost.

For each collector surface about equal to the pool surface area a 10 degree F temperature rise in the average temperature of the pool can be expected.

More sophisticated and more efficient solar collectors can be designed and used but their cost can hardly be justified unless these same collectors are also used for heating the house, heating the water, and possibly solar air conditioning.

The collectors for swimming pool heating can be made of plastic, however, it must be realized that the life of such systems will be shorter than the properly designed metallic equivalents.

# Simple Method for Evaluating Performance of a Flat Plate Air Collector

By Paul Shippee

It is necessary to introduce and define the notion of efficiency because performance of a certain size unit is measured by efficiency. In general, Output divided by Input gives a ratio which multiplied by 100 is the efficiency in percent. Output is always less so it is necessary to locate and identify the losses. For a solar collector efficiency is the fraction of heat delivered to use that actually fell on the collector from the sun. The losses are optical reflection at the air glass interfaces, convection to and conduction through the structural parts including transfer ducts (which is recoverable if they are indoors), air leaks and infiltration, and radiation to the glass from the heated components. If the wind is blowing the thermal losses through the structure will be increased greatly by the convection loss from the outer surfaces to the surrounding air.

For an air collector used to heat up rocks or water in storage it is useful to think of the performance as the capacity to heat up these materials. Using this approach I constructed a collector, tilted it toward the sun and mounted an insulated box with a door in it on top. Coffee cans with tops were filled with water and spaced one inch apart on open shelves in the box. A door was provided for measuring the temperature of storage water inside. The heated air passes upward from the collector into the box. A baffle makes it go to the top before moving sideways to the colder cans. Cooled by the colder cans the air drops and enters the return duct, a separate passage built into the collector, displacing the air being heated by the blackened aluminum absorber plate. It is a closed free convective air loop with the same air being sequentially heated by the radiation absorber and then cooled by the water cans.

Eventually the water gets heated and the free convection system is rendered less and less efficient as the temperature differential is reduced and the circulating process slows down.

The temperature of the absorber plate is inversely proportional to the velocity of the air moving by it and directly proportional to the temperature of the air returning to it from storage. The higher the temperature of the absorber plate the higher the thermal losses mentioned above. The lower the temperature of the absorber plate the lower the temperature of the heat delivered to storage. A balanced mode of operation lies within a narrow range somewhere between these two extremes.

Around noon two things happen. The air velocity slows down due to higher temperature in storage and the temperature of the air returning to the absorber from storage gets higher and

**Solar air collector - Water Storage - Working Model**

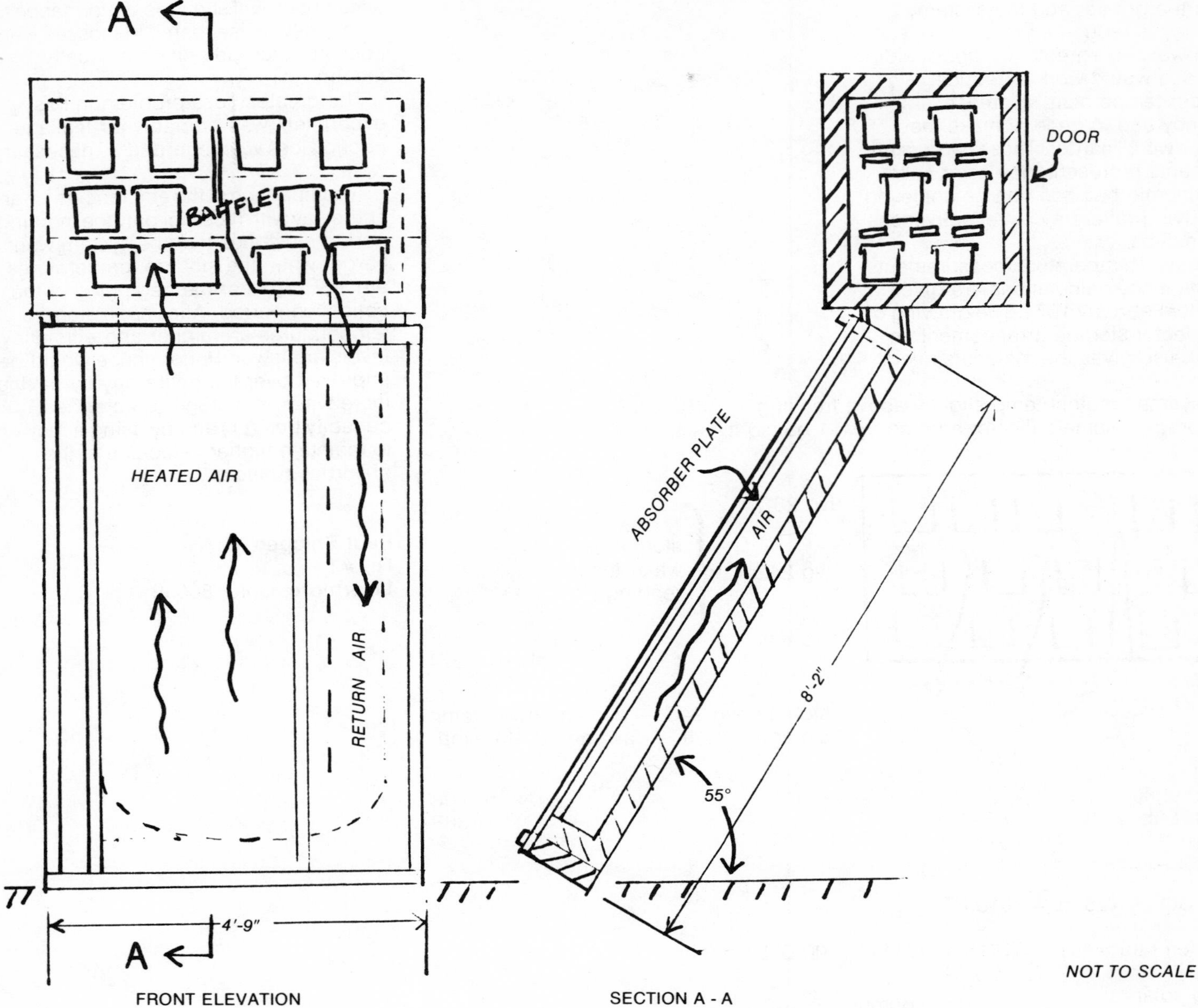


FRONT ELEVATION

SECTION A - A

*NOT TO SCALE*

NOTES:

Storage Box: ½" *Plywood on 2 x 4 frame.*
*3" Polyurethane insulation all sides*
*3 lb. coffee cans with plastic tops mounted on open slat shelves rear side hinged as a door.*

Collector Box: Framed with 2 x 10 lumber, backside ¾" *boards.*
*4" Polyurethane scraps sides and bottom insulation.*
*Absorber plate: Corrugated aluminum with barbecue black paint.*
*Glazing:*
*Outer, 5 mil mylar exposure life 6-10 months.*
*Inner, 2 mil mylar stapled to back of wood frame.*
*These films 1" apart, after 2 months they are stuck together near bottom.*
*Wind flap excessive at all times in any wind.*
*Exposed 2 mil film crumples like old leaf after 2 months.*

higher also due to the higher temperature in storage. Both of these things contribute to higher temperatures at the absorber plate and so the collector losses are greater and the system efficiency is reduced.

A blower and a greater volume of water in storage would work to reduce absorber temperature, increase efficiency and in general make the system "work" harder. These two system increments represent increased costs. An economic balance can be achieved by studying different cost/effectiveness relationships.

Results of temperature measurements and efficiency calculations are given below for February 1974. The drawing of the collector-storage arrangement on page 2 also gives the materials used.

—daytime ambient temp slightly above freezing
—storage volume/collector area approx. 1 gal/sq ft

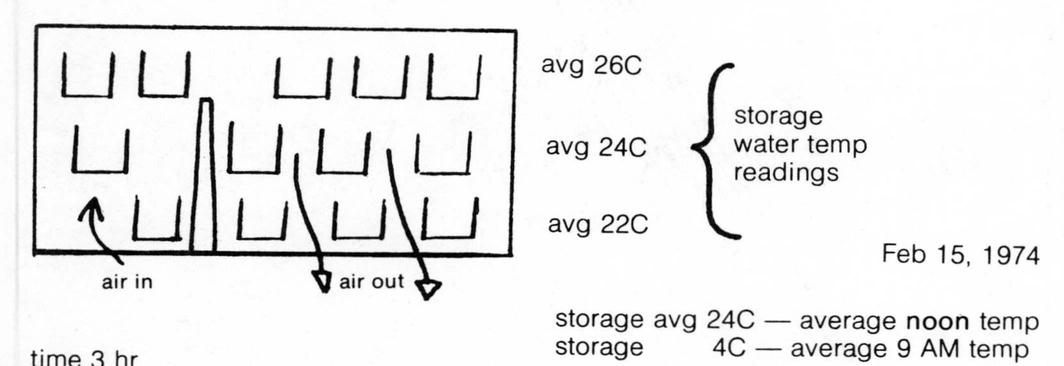


Feb 15, 1974

storage avg 24C — average **noon** temp
storage 4C — average 9 AM temp

time 3 hr
coll.
area 15 sq ft
water 128 lb

T = 20C
T = 36F

BTU = 36F × 128 lb = 4610 BTU

Collection rate =4610 BTU/15 sq ft/3 hr 100 BTU/sq ft/hr

For avg solar insolation 225 BTU/sq ft/hr

$$\text{Efficiency} = \frac{\text{output}}{\text{input}} = \frac{100}{225} = 44.5\%$$

a week later similar measurements for **entire** day give

$\Delta T = 31C - 2C = 29C = 52F$

time = 7 hr

BTU = 52F × 128 lb = 6650 BTU

Collection rate = 6650 BTU/15 sq ft/7 hr = 63 BTU/sq ft/hr

$$\text{Efficiency} = \frac{63}{225} = 28\%$$

The losses from the collector box are not counted because it is only wanted to know the capacity of a particular collector for heating the storage material. The losses are accounted for the final results.

The losses from the storage box are neglected because during daytime exposure to sunny weather the losses are minimal. In other words it is assumed for this simple method that all the heat delivered to storage stays there. This assumption will show the performance to be slightly worse than if the losses were counted. Storage units are usually indoors.

The greatest loss problem in this experiment was probably air leakage. Friction loss was regarded as necessary work.

A comparison of the efficiencies for an entire day with those for only the morning period illustrates that the system works harder when the storage temperatures are low, i.e. near the beginning of the collection period. After noon storage temperatures are higher and efficiency levels are lower. Performance would be improved over the entire day by having more water in storage (greater heat capacity per $\Delta$T) and by using a blower to maintain higher velocities at the absorber surface.

**Paul Shippee**
**Rt. #1**
**Livermore, Colo. 80536**

# A Solar Heated Bath

By Tyler Volk

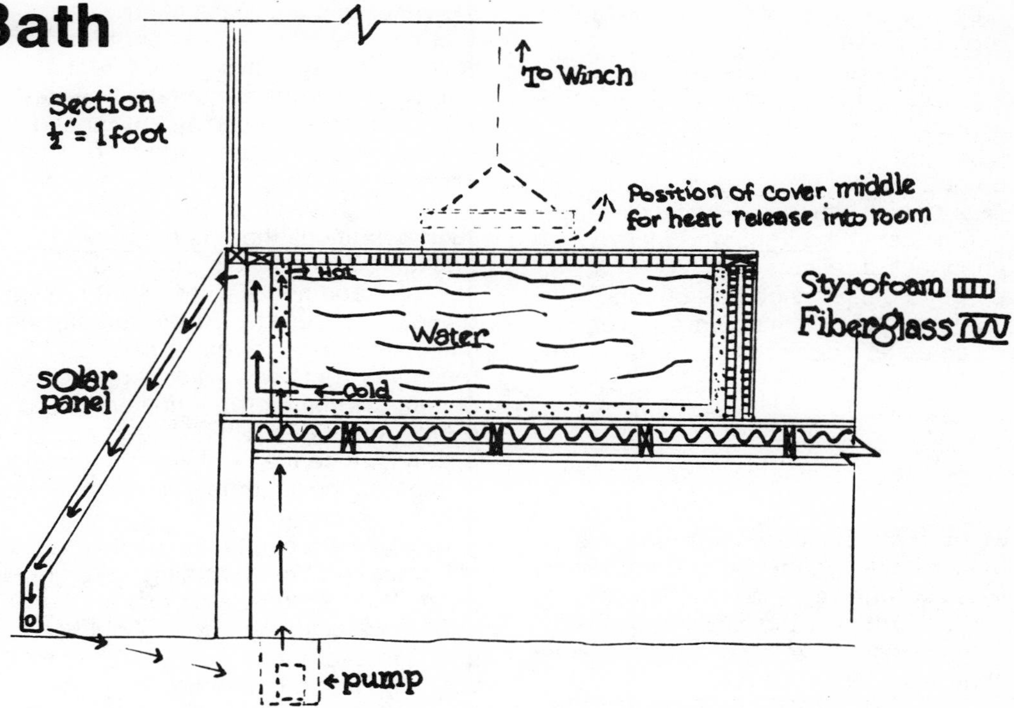


In the fall of 1973, three of us went looking for a house to rent for winter. During the search we met and became friendly with Mrs. Minna Peyser, an unusual woman who was very open to new ideas. We discovered that some of our dreams overlapped with hers, and after several months of designing, researching, and testing, we have begun remodeling her home.

The project includes, among other things, the construction of a large whirlpool bath which will be heated with a 150 sq. ft. flat plate solar collecter. We found that we will be able to heat the pool water to bathing temperature (100°F.) and supply additional heat to the house, cutting down on gas bills.

We first constructed a test panel, to test the system's capabilities. This was monitored carefully from January to February, 1974. It was about one tenth the size of the system for the Peyser house.

Test Panel Info

description:
a submersible pump in a 50 gallon insulated drum pumps water to the top of the panel, it flows down and returns by gravity to the drum. 300 pounds of water are in the drum

highest temp. reached:
115°F. in drum
124°F. directly from panel
(the water off the panel averaged 10° higher than the water into the panel)

collection efficiency:
heat collection at a low temperature range (40°-90°) collected at a 50% higher efficiency than collection at a high temperature range (90°-110°)

low range: 1000 BTUs/sq. ft./day
high range: 750 BTUs/sq. ft./day

**The SYSTEM**

The actual system is different from the test panel in that the water is allowed to flow from the pool by gravity, down the panel, and into a container where it is pumped back into the pool. This was the most convenient placement of a solar panel for this particular house. The pool is insulated all around with two layers of insulation and an air space, including an insulated cover. The cover can be completely lifted by a winch when the pool will be used, or, the pool's center insulation can be lifted, allowing the heat to move more quickly from water to room air. The pool holds about 400 gallons of water.

We are two young designers who are interested in integrated energy design. We have some knowledge and skills in alternative energy sources, domes, close packing polyhedral buildings, and integrated energy design.


Tyler Volk, Mark Hughes
RD 1, Hitchcock Hill Rd.,
Mahopac, N.Y. 10541


# A Soap Bubble Heat Collector

By Steve Baer

I enjoy playing with bubbles very much. I have often tried to use them in different projects. Bubbles are such perfect things. Perfect spheres, if they are alone, perfect hemispheres, if they are against a wet surface. And then, a foam of bubbles adjusting to each other in a forest of perfect 120° angles linking dozens of different curvatures is, perhaps, even more marvelous. You see the bubble form, so superb, alone, a kind of prima donna in space is ready to adjust to other bubbles and form beautiful societies of bubbles. There are excellent books about soap bubbles. CV Boy's **Soap Bubbles**, Dover Press, is a lovely book. Darcy Thompson has many interesting things to say about bubbles in **On Growth and Form** and, more recently, Peter Steven's **Patterns in Nature**, Little Brown, shows a proper love and appreciation for the bubble. Why should anyone go to school when he or she can stay home with a pan of water, some soap and a straw?

Bubbles can even stop you from smoking. Blow some smoke bubbles—mixed with clear bubbles, they look like puss infected cells—when they pop, the collapsing films are brown and gummy, unmistakably the same stuff that collects in your mouth and lungs.

My idea for a bubble solar collector was to use soap bubbles as a clear insulating material in the space between the collecting surface and a pane of glass. The solar collector designer's great problem is that as he exposes his black surface to collect sunlight, he simultaneously exposes it to lose heat. It is very important to design something that will let the sun in, but will not let the heat out. This seemed a wonderful place for the soap bubble. Why not fill a space with a clear foam? The sun can pass through, but when the heat tries to leave it finds itself trapped by the same material sold commercially to insulate freezers, ice chests and buildings—foam.

I got a bicycle pump, and an inner tube as an air reservoir, some surgical tubing and a valve and found that bubbles were even more interesting if they flowed endlessly and effortlessly from such a simple mechanism. Later I replaced the bicycle pump with a tiny air compressor used for spray painting and the small inner tube with a gigantic tube from an earth mover on which we had floated down the Rio Grande. You lay the air nozzle in a pan of soapy water and turn it on. A small nozzle blows small bubbles, a large nozzle blows large bubbles. If the bubbles are exposed to the outside dry air, they soon pop, but if the bubbles are blown into a compartment, the air in the compartment becomes saturated and the bubbles live for hours and hours.

M. Kudret Selcuk published an article in the Solar Energy Journal, Vol. #8, no. 2. 1964, showing the excellent performance of a clear plastic foam, Styrocel, as a glazing material for solar collectors. Why couldn't soap bubbles replace plastic?

In solar collector design there are many variables to play with. Every time you add a layer of glass to a solar collector, you help it retain the heat it has gained, but you simultaneously cut down its supply of heat because the new layer of glass absorbs light, and more important, it reflects light. A perpendicular ray going from one medium to another medium has a portion reflected as it crosses the interface. This portion equals

$$\left(\frac{n-1}{n+1}\right)^2$$

where n equals the refractive index between the two media. The refractive index between air and water is about 1⅓ and between air and glass about 1½, then the refractive index between water and glass is about 1⅛.

To explain the 1⅛, if you pick 12 to name the velocity of light in air, it then follows that 8 and 9 are the velocities of light in glass and water respectively. Between the two substances, then, the index of refraction is 1⅛. The reflection from air to glass is

$$\left(\frac{1\ 1/2-1}{1\ 1/2+1}\right)^2 = \left(\frac{1/2}{5/2}\right)^2 = 1/25 = 4\%.$$

The reflection from air to water is

$$\left(\frac{1\ 1/3-1}{1\ 1/3+1}\right)^2 = \left(\frac{1/3}{4/3}\right)^2 = \frac{1}{49} = 2\%.$$

The reflection from water to glass is

$$\left(\frac{1\ 1/8-1}{1\ 1/8+1}\right)^2 = \left(\frac{1/8}{17/8}\right)^2 = 1/289 = 1/3\%.$$

The total reflection going through a dry pane is about 8% and through a wet pane about 4⅔%. And the reflection going through a layer of water is 4%.

It appears that if we keep our glass wet we can cut down the reflective losses by a great deal. I can have two wet panes in the place of one dry pane and have hardly any more reflected light. The equations show that each bubble wall is also approximately ½ as reflective as a pane of glass.

But we are misusing some of our equations. These equations tell us the reflection of light if the materials are relatively thick compared to a wave length of light and it turns out that bubbles are often as thin as wavelengths of visible light. One fifty thousandths of an inch. With films as thin as this our equations are no longer accurate. Soap bubble films in great numbers reflect very little light. I filled trays partly full of soapy water, placed black mats under the water so that light would be absorbed, then put panes of glass over the trays. I raced one tray against another. One tray had the space between the water and the glass filled with bubbles, the other had no bubbles. The water in the two trays gained heat at approximately the same rate. The tray without bubbles pulled slightly ahead of the bubble tray at the beginning. This made sense. The water in the two trays was close to the ambient temperature. In such conditions there is no point in providing insulation to protect a collector against heat loss for it has no place to lose heat to. Any added glazing makes the collector pay a price by reflecting light away without giving benefit. Swimming pool heaters used during mild times of year need no glazing, since the pool water is usually so close to the air temperature that glass offers no advantage.

The difference in the respective rates of heat gain at the begining were not great. This demonstrated that all the bubble walls through which light had to pass (an average of 6 to 10) really took very little toll in reflection. After a time, the two trays reached equilibrium with the bubble tray slightly warmer than the tray without the bubbles. There was a marked difference in the glass temperature of the two trays. The bubble tray's glass was noticeably cooler.

Why wasn't there a dramatic difference? Here I'd found an insulating material that stopped the convection of air by breaking it into tiny parcels and which was extraordinarily clear, yet it seemed to make hardly any difference. The problem finally became apparent. All the time I was working with bubbles I was counting on their selective transmissivity. Water is like glass—it allows visible light through, but blocks infra red radiation. (Thin films of plastic are transparent to infra red—test this by getting a stove or a hot plate hot and first passing a pane of glass and then a sheet of plastic between the hot plate and your face. The glass blocks the heat completely, but the thin plastic hardly at all.) Evidently the bubbles are so thin, a small fraction of the infra red wave lengths, that the infra red goes right through them. Radiant transfer is such an important part of the heat loss of a collector that my failure to halt it left the collector poorly insulated.

Another possible cause for the poor performance of the bubbles was the continual flow of new air into the bubble space.

The air arrived dry and cool and became hot and wet as it filled bubbles and then squeezed out an exhaust hole opposite the bubble pan to float away in the wind.

Often failures such as the bubble collector turn out to be very satisfactory in other strange ways. As a tree that has its main trunk damaged begins to bud at unlikely points along the trunk base, so, when faced with a project that doesn't work, you begin to tinker with interesting sidelights. I believe that to do good design work you must love the materials. When they do not cooperate in their first arrangement, by giving you exactly what you want, you must realize that you probably have not understood the materials and processes before your eyes.

Here the bubbles migrated slowly from one end of the chamber to the other. It looked like some kind of magic fleece. As the bubbles age, they thin and interference patterns cause them to reflect all the colors of the rainbow. The foam winks and twinkles as it slowly moves behind the wet glass.

I am convinced that someone someday will find a way to use these beautiful forms in insulating walls. The entire walls are alive with migrating, refracting bubbles. Who cares whether or not they collect solar energy!

Another use of soap bubbles:

You may have noticed while washing dishes that if you take a hot wet glass and place it on a slightly wet drain board, as it cools it blows bubbles inside itself around the rim. Here you have a demonstration of the contraction of air as it cools. The volume of a gas is proportional to its absolute temperature. The volume of bubbles blown inside a cooling glass seems a bit extreme, if you study the small swing in temperature the glass goes through. Something else is happening in the glass. The water vapor is condensing and this adds dramatically to the shrinkage in the glass. One can feel the reverse of this by partly filling a bottle with hot water and shaking it. Evidently the water sitting in the bottom fails to humidify the air completely, for when the bottle is shaken you can feel the surge of pressure against your hand as the gas expands.

A year after my first experiments with soap bubble collectors, Day Chahroudi arrived in Corrales and began a number of experiments using bubbles and boiling liquids. Day's extensive knowledge of chemistry plus an unspoiled love of such beautiful things as bubbles greatly added to the excitement. Day has now worked with many clear insulating membranes going on to use selective reflective films and different geometries within quilted fabrics controlled by air pressure.

The events that take place with air and water at different temperatures are fascinating to observe. I would recommend getting a clear bottle—a Coors beer quart bottle with its twist on aluminum cap sealed by a polyethelene gasket is a nice bottle and—filling it almost full of water and admiring it. When you touch a bottle filled with water you become aware of its insistance on maintaining its own temperature. It is a calm authority you feel as a cool bottle steadily takes heat from your hand or a warm bottle supplies it with heat. Water has a stable disposition this way. It is able to hold as much heat as an equal volume of metal and twice as much as stone or masonry. Picking up an old bottle from the trash, cleaning it and filling it with water and playing with it is exciting in another way. It is evidence of the beauty and inexpensiveness of machine made products. The can is not nearly so satisfactory unless it is one of the new aluminum cans, because always in the back of your mind is the thought of rust and corrosion. It is interesting to watch the bubbles in a partly filled bottle always finding the highest position, always rounding its edges where it meets the glass, always maintaining the same volume. Spilled water is such a completely different substance that containers are true marvels. If you buy a thermometer you can take the temperature of the water in your bottle. A quart bottle holds two pounds and, therefore, a 1°F rise in temperature indicates a gain of 2 Btu. A thermometer is essential. Take the temperature of everything, the dirt, the air, your shower water (both at the showerhead and the drain), your coffee, your soup, your own urine as you urinate. A thermometer can be placed outside the windshield of a car so that you can read it through the glass. Small clips siliconed to the outside of the windshield will hold a thermometer (with a tin shield for the bulb) in place even at high speeds. Such a thermometer acquaints you with the variations in temperature associated with different microclimates. Early in the morning valley floors are noticeably colder than sidehills, and, more dramatic, as a drive into or out of a city during the evening. The buildings store heat and there is commonly a 10°F difference in temperature between downtown and the suburbs. The zone of changing temperature on the edge of a downtown is called a "heat cliff" by micro meteorologists.

I would recommend to anyone interested in solar energy that he play extensively with simple things such as bubbles, bottles of water, stones, pieces of metal and thermometers. Our society's approach to problems today is often to leap frog the simple, obvious solution and land in the midst of computer programs and complicated machines where they really are not needed.

Steve Baer
Zomeworks Corporation
PO Box 712
Albuquerque, N.M. 87103

---

**Solar Still Plans**

Brace Do-it-Yourself Leaflet L-I
"How to make a Solar Still (Plastic covered)"
$.75

Brace Technical Report T 17
"Simple Solar Still for the Production of Distilled Water"
$.75

Brace Technical Report T 58
"Plans for a Glass and Concrete Solar Still"
$3.50

The plans above are all available from:
Brace Research Institute
Macdonald College of McGill University
Ste. Ann de Bellevue 800
Québec, Canada

Sunwater makes small (1 to 2 gallons per day) solar stills for people who are tired of paying good money for bottled water. A wide variet of options is available.

information from:
Sunwater Co.
1112 Pioneer Way
El Cajon, Ca. 92020

The following manufacturers of solar equipment have brochures or information available on their products.

Fun & Frolic, Inc.
P.O. Box 277
Madison Heights, Michigan 48071

*Makers of "Solerator," a flexible plastic solar water heater for pool or water heating.*

Sun Source of Daylin, Inc.
9606 Santa Monica Blvd.
Beverly Hills, Calif. 90210

*Distributors of "Miromit" solar hot water heaters, made in Israel and used widely in several countries.*

Fred Rice Productions, Inc.
6313 Peach Avenue
Van Nuys, Calif. 91401

*Two brochures on solar energy as applied to mobile homes ($1.00 each) plus a free brochure on the "SAV" solar hot water heater.*

Tranter Manufacturing, Inc.
735 E. Hazel Street
Lansing, Michigan 48909

*Makes the "Econocoil" solar hot water collector.*

Sunworks, Inc.
669 Boston Post Road
Guilford, Conn. 06437

*Makes and markets two different types of solar hot water heater. Technical data and drawings, $1.00.*

Kalwall Corp.
1111 Candia Road
P.O. Box 237
Manchester, New Hampshire 03105

*Makes "Sun-Lite," a clear flexible fiberglass film for covering solar collectors.*

Energy Corp.
481 Tropicana Road
Las Vegas, Nevada 89109

*Makes two types of aluminum solar hot water heaters.*

# Utilization of Sun and Sky Radiation For Heating and Cooling of Buildings

By John I. Yellott
*Fellow ASHRAE*
*Chairman, ASHRAE Technical Committee–6.7*
*Solar Energy Utilization*


There are three technological processes by which solar energy can be utilized: (a) heliochemical; (b) helioelectrical; (c) heliothermal. The first process, through photosynthesis, maintains life on this planet; the second process, using photovoltaic converters, provides power for all of the communication satellites and will ultimately be very valuable for terrestrial applications. The third process is the primary subject of this chapter, since this can be used to provide much of the thermal energy needed for space heating and cooling and domestic hot water heating.

## SOLAR HEAT COLLECTION

Solar radiation data may be used to estimate how much energy is likely to be available to be collected at any specific location, date and time of day by either a concentrating device, which can use only the direct rays of the sun, or a flat-plate collector which can use both the direct and diffuse radiation. Since the temperatures which are needed for space heating and cooling are quite moderate, not exceeding 250°F even for absorption refrigeration, they can be attained by carefully designed flat-plate collectors and so the fundamental principles of these simple but effective devices will be discussed in some detail in the following sections.

A flat-plate collector, as its name implies, generally consists of the five components shown in Fig. 1. [(1) Glazing, which may be one or more sheets of glass or a diathermanous (radiation-transmitting) plastic film or sheet. (2) Tubes or fins for conducting or directing the heat-transfer fluid from the inlet duct or header (2-in) to the outlet (2-out). (3) Plate, generally metallic, which may be flat, corrugated or grooved, to which the tubes or fins are attached in some manner which produces a good thermal bond. (4) Insulation, which minimizes downward heat loss from the plate. (5) Container or casing, which surrounds the foregoing components and keeps them free from dust, moisture, etc.]

During the past century, flat-plate collectors have been constructed from many different materials and in a very wide variety of designs. They have been used to heat water, water plus an anti-freeze additive such as ethylene glycol, water plus ammonia or other refrigerants, fluorinated hydrocarbons, air and other gases. The major objective has been to collect as much solar radiation as possible, at the highest attainable temperature, for the lowest possible investment in labor and materials. When these goals are attained, the collector should also have an effective life of many years, despite the adverse effects of the sun's ultraviolet radiation, corrosion or clogging due to acidity, alkalinity or hardness of the heat-transfer fluid, freezing and air-binding in the case of water or deposition of dust and moisture in the case of air, and breakage of the glazing due to thermal expansion, hail or other causes.

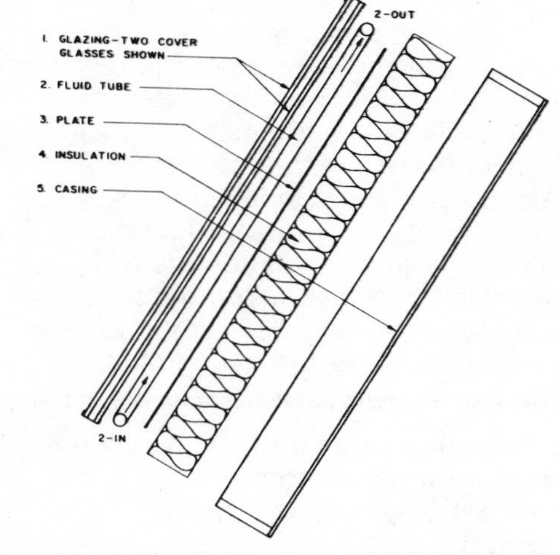


Fig. 1 Exploded cross-section through typical solar water heater

## GLAZING MATERIALS

Glass has been the principal material used to glaze solar collectors because it has the highly desirable property of transmitting as much as 90% of the incoming shortwave solar radiation, while virtually none of the longwave radiation emitted by the flat plate can escape outward by transmission. Glass of low iron content[1] has a relatively high transmittance (approximately 0.85 to 0.90 at normal incidence) for the solar spectrum from 0.30 to 3.0$\mu$, but its transmittance is essentially zero for the longwave thermal radiation which is emitted by sun-heated surfaces (3.0 to 50$\mu$).

Plastic films and sheets also possess high shortwave transmittance but, because most of the usable varieties possess transmission bands in the middle of the thermal radiation spectrum, they may have longwave transmittances as high as 0.40[1,2].

Plastics are also generally limited in the temperatures which they can sustain without undergoing dimensional changes and only a few of the varieties now available can withstand the sun's ultraviolet radiation for long periods of time. They possess the advantage of being able to withstand hail and other stones and, in the form of thin films, they are completely flexible.

The glass which is generally used in solar collectors may be either single-strength (0.085 to 0.100 in. thick) or double-strength (0.115 to 0.133 in.) and the commercially available grades of window or green-house glass will have normal incidence transmittances of about 0.87 and 0.85 respectively. For direct radiation, the transmittance varies to a marked extent with the angle of incidence, as shown by Table 1 which gives transmittances for single and double glazing using double-strength clear window glass.[3]

For clear glass such as that used for solar collectors, the 4% reflectance from each glass-air interface is the most important factor in reducing transmission, although about 3% gain in transmittance can be obtained through the use of "water white" glass. Anti-reflection coatings of the kind used for camera and telescope lenses can also make significant improvement in transmission, but the cost of the processes presently available is prohibitively high.

The effect of dirt and dust on collector glazing is surprisingly small, and the cleansing effect of occasional rain seems to be adequate to maintain the transmittance within 2 to 4% of its maximum value[1,4].

The purpose of the glazing is to admit as much solar radiation as possible and to reduce the upward loss of heat to the lowest attainable value. Glass is virtually opaque to the longwave radiation which is emitted by the collector plate but the absorption of that radiation causes the glass temperature to rise and thus to lose heat to the surrounding atmosphere. This type of heat loss can be reduced by using an infrared-reflecting coating on the under-side of the glass[1] but such coatings are costly and they also reduce the effective transmittance of the glass for solar radiation by as much as 10%.

In addition to serving as a heat-trap by admitting shortwave solar radiation and retaining longwave thermal radiation, the glazing also reduces heat loss by convection. The insulating effects of the glazing is enhanced by the use of several sheets of glass, or glass plus

**Table I. Variation with incident angle of transmittance for single and double glazing and absorbance for flat black paint.**

| Incident Angle, degrees | 0 | 10 | 20 | 30 | 40 | 50 | 60 | 70 | 80 | 90 |
|---|---|---|---|---|---|---|---|---|---|---|
| Transmittance | | | | | | | | | | |
| Single glazing | 0.87 | 0.87 | 0.87 | 0.87 | 0.86 | 0.84 | 0.79 | 0.68 | 0.42 | 0.00 |
| Double glazing | 0.77 | 0.77 | 0.77 | 0.76 | 0.75 | 0.73 | 0.67 | 0.53 | 0.25 | 0.00 |
| *Absorptance for flat black paint | 0.96 | 0.96 | 0.96 | 0.95 | 0.94 | 0.92 | 0.88 | 0.82 | 0.67 | 0.00 |

*Adapted from [1].

plastic[4]. The upward heat loss may be expressed by:

$$Qup = Acp \times Ucp \times (tp-to) \text{ Btuh} \quad (17)$$

where Qup designates the heat loss upward from the collector plate of area Acp sq ft, Uup is the upward heat loss coefficient in Btuh/ft² F, and tp and to are the temperatures of the collector plate and the ambient air respectively.

## COLLECTOR PLATES

The primary function of the collector plate is to absorb as much as possible of the radiation reaching it through the glazing, to lose as little heat possible upward to the atmosphere and downward through the back of the container, and to transfer the retained heat to the transport fluid.The absorptance of the collector surface for shortwave solar radiation depends upon the nature and color of the coating and upon the incident angle, as shown in Table 1 for a typical flat black paint.

Prior to 1955, flat black oil-based paint was the universal choice for coating solar collectors. The publication of papers by Tabor of Israel and Gier and Dunkle of the U.S.[4,5,6] led to awareness of the fact that there is virtually no overlapping of the shortwave solar spectrum (0.3 to 3.0μ) and the longwave thermal spectrum (3.0 to 30μ). By suitable electrolytic or chemical treatments, it is possible to produce surfaces which have high values of solar radiation absorbtance $\alpha$, and low values of longwave emittance, $e_S$[5,6,7,8,9]. Essentially, such "selective surfaces" consist of a very thin upper layer which is highly absorbant to shortwave solar radiation, but relatively transparent to longwave thermal radiation. The substrate must have a high reflectance and a low emittance for longwave radiation.

Among the selective surfaces which have demonstrated their ability to retain their desirable properties after long exposure to intense sunshine are those produced by Tabor's electrolytic processes and used in the mass production of solar water heaters in Israel, with galvanized sheet steel as the plate material. Solar absorptances in the range of 0.92 and longwave emittance as low as 0.10 characterize the Tabor surfaces.

Daniels[7] gave a comprehensive analysis of the chemical aspects of producing selective surfaces and Christie[9] gives details of current Australian practices in this very important field.

Selective surfaces are particularly important when the collector surface temperature is much higher than the ambient air temperature. The references listed above are only a few of the very large number of publications which have appeared on this subject, many as a result of the use of selective surfaces in the space program.[10]

If the objective of the collector is to heat liquids, tubes must be integral with or attached to the plate with a good thermal bond. Beginning with the classic work of Hottel and Woertz[11] at M.I.T. and continuing with the studies made by Whillier and Erway, many investigations, both theoretical and experimental, have been made into the details of collector design and performance. The method of attaching the tubes to the collector plate has challenged the ingenuity of designers in all parts of the world. The major problem is to obtain a good thermal bond without incurring excessive costs for labors or materials.

Materials most frequently used for collector plates, in decreasing order of cost and thermal conductivity, are copper, aluminum and steel. If the entire collector area is swept by the heat transfer fluid, the conductivity of the material becomes unimportant. Whillier has studied the effect of bond conductance[12] and has concluded that steel pipes are just as good as copper if the bond conductance between tube and plate is good. Bond conductance can range from a high of 1,000 Btuh/ft/F for a securely soldered tube to a low of 3.2 for a poorly clamped or badly soldered tube. Bonded plates with integral tubes are among the best alternatives as far as performance is concerned, but they require mass production facilities.

Fig. 2, adapted from [13] and other sources, shows a few of the very large number of solar water and air heaters which have been used in the past with varying degrees of success. It will be a very ingenious designer who can devise an entirely new type of heat absorber which has not been tried previously by someone working in Australia[14], France, India, Japan[15], South Africa[13] or the U.S.

In Fig. 1 (A) shows a bonded sheet design, in which the tubes are integral with the sheet, thus guaranteeing a good thermal connection between the plate and the tubes. This process is widely used commercially for producing radiators and other heat exchangers. (B) and (C) show conventional fluid heaters with the tubes soldered or otherwise fastened to upper or lower surfaces of metal sheets. Clips, clamps, twisted wires, thermal cement and many other devices have been tried with varying

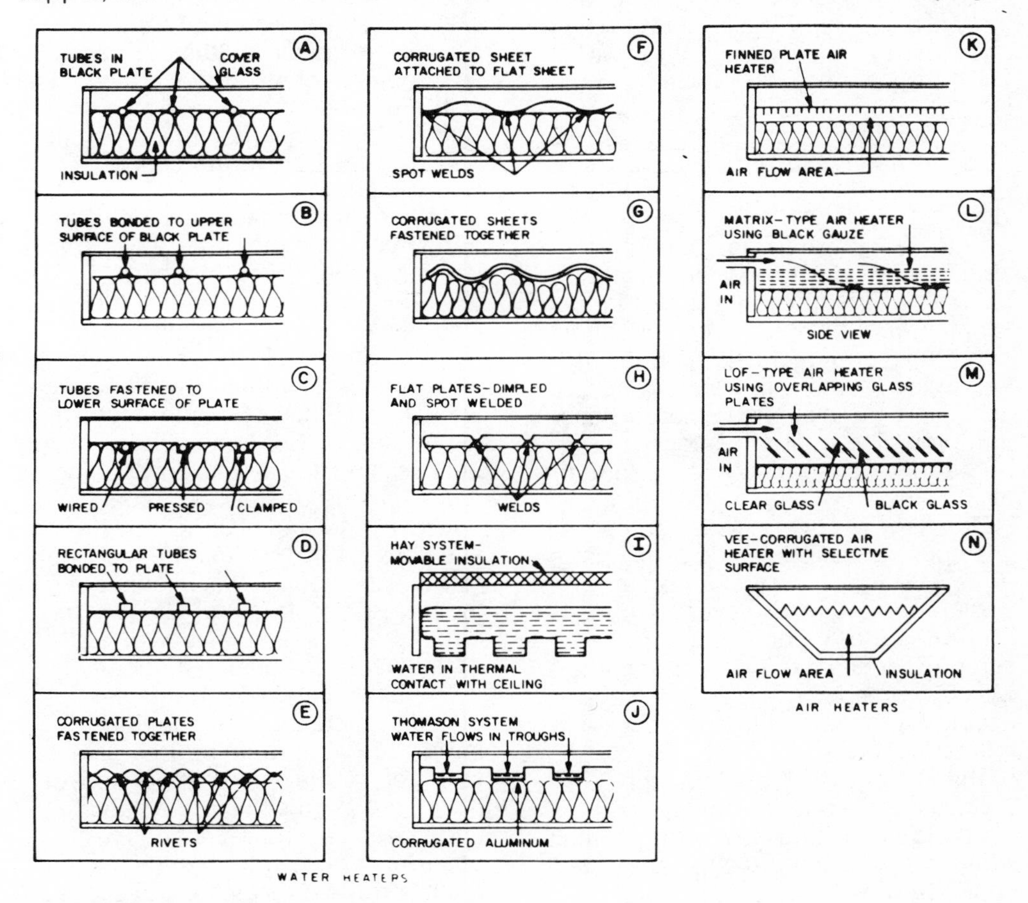


Fig. 2 Variations of solar water and air heaters

degrees of success. (D) shows the use of tubing with rectangular rather than circular cross-section, to obtain more contact between the tube and the plate. Mechanical pressure, thermal cement or brazing may be used to make the actual assembly.

(E), (F) and (G) show different ways in which galvanized steel sheets have been fastened together to make water-tight containers with individual fluid passages. (H) shows the concept of using parallel sheets of copper, aluminum or galvanized steel which are dimpled and fastened together at intervals by spot-welding or riveting. All of the non-tubular types are limited in the pressures which they can sustain and they are not, in general, suited for use with the high line pressures used in the U.S.

Fig. 2 (I) shows the horizontal roof ponds used in the Hay system[16] in which water is contained in transparent plastic bags supported by a metal ceiling, with movable horizontal insulation above the water surface. For summer cooling, when additional heat dissipation must be obtained by evaporation, the bags may be flooded to provide an uncovered surface. The insulation covers the pond during summer days and winter nights.

Fig. 2(J) shows the Thomason corrugated aluminum channel collector which is used with steeply-pitched south-facing roofs[17 18]. Water is distributed to the channels by perforated copper piping running along the peak of the roof and collected at the bottom of the channels by a galvanized trough which leads the warmed water to the basement storage tank.

The heating of air or other gases can be done readily with flat-plate collectors, particularly if some type of extended surface (Fig. 2-K) can be used to overcome the low values of the heat transfer coefficient between metal and air. Matrix-type material (L), overlapped sheets (M) of glass or aluminum and many other approaches have been used to provide a means of trapping and absorbing the incoming solar radiation and providing a large contact area between the absorbing material and the air.

Breakage of cover glasses due to hail or stones can be prevented by the use of ½ in. wire mesh, supported on an angle-iron framework several inches above the glass. The use of such a screen reduces the effective absorber area by about 15%, and so the total area should be increased to compensate for this reduction.

The casing which contains the collector plate and insulation and supports the glazing may be made with a wooden frame and a moisture-resistant headboard backing or it may be constructed of sheet metal, carefully painted to resist rust. Asbestos cement is another material which has been used successfully, but it is relatively heavy and it is breakable. Plastic cases reinforced with glass fibers have the advantage of being weatherproof and they are well adapted to mass production. The cover glass must in any case be firmly secured with suitable weather-resistant gaskets. Sheets of glass larger than about 4 ft square become unwieldy and difficult to handle, so large collectors are generally glazed with a number of smaller lights. Aluminum extrusions and neoprene gaskets may be used to connect the individual lights.

If the collection area is more than adequate, as in the case of the French residences in which the entire south-facing concrete walls of the structures are available, all that is needed is a coat of dark paint on the wall and a space of about 1 in. width between the wall and the glazing through which the air can flow. If a metallic collector plate is used, the air can flow behind the plate, leaving a "dead air" space between plate and glass to provide some anti-convection insulating effect. Metal ribs can then be used at intervals to give the necessary structural support, to direct the air stream and to give additional heat transfer surface.

The selectivity of an air heater collector plate can be markedly improved by corrugating it to give a series of parallel vee-shaped grooves[19]. Direct radiation striking the vees undergoes several reflections, with absorption occurring at each surface. Such corrugated plates are directionally selective and, when mounted with the proper orientation, they display a solar absorptance during most of the year which is considerably higher than the absorptance of the flat sheet from which they are made. The increase in emittance as compared with the flat sheet is relatively small. The vee-corrugated sheets provide additional heat transfer surface, as well as improving the absorptance.

At the 1961 U.N. Conference on New Sources of Energy, G. Francia[20] proposed the use of transparent honeycomb structures between the surface of a flat-plate collector and its cover glass. The honeycomb suppresses convection to a very marked degree and the longwave radiation loss is also significantly diminished. More recent studies[21] have verified his findings, and research is continuing in this field to find honeycomb materials which are sufficiently inexpensive and durable to be used on a large scale.

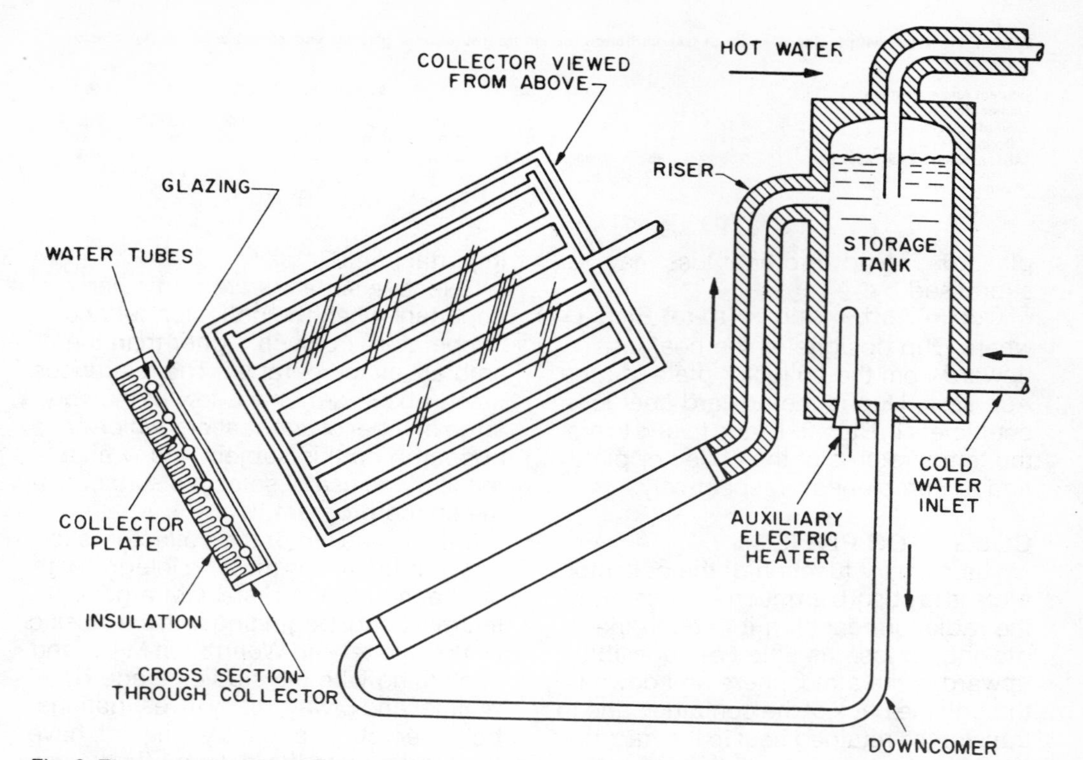

Fig. 3 Thermosyphon water heater showing details of storage tank and heat collector. (Courtesy HPAC)

## APPLICATIONS OF SOLAR RADIATION COLLECTORS

**Thermosyphon heaters.** The most widely-used heliothermal device is undoubtedly the thermosyphon solar water heater, Fig. 3. A flat-plate collector, generally one of the tubed varieties shown on Fig. 2, is mounted so that it faces south with a tilt (from the horizontal) equal to the latitude plus or minus 10 or 20 deg, depending upon the season of year when maximum output is needed.

An insulated storage tank, 40 to 60 gallons in capacity, is mounted so that its bottom is at least 2 ft above the top of the collector and an insulated downcomers carries cold water to the header at the bottom of the collector. Another insulated pipe connects the upper header to an inlet tap approximately two-thirds of the way up towards the top of the tank. The system is filled with cold water and when the sun shines on the collector, the water in the tubes is heated. The warm water flows slowly upward through the tubes and the riser into the storage tank, and this process continues as long as the sun's radiation is strong enough to keep the collector plate hot. Fig. 4 shows the results of a typical day-long test of the unit shown in Fig. 3.

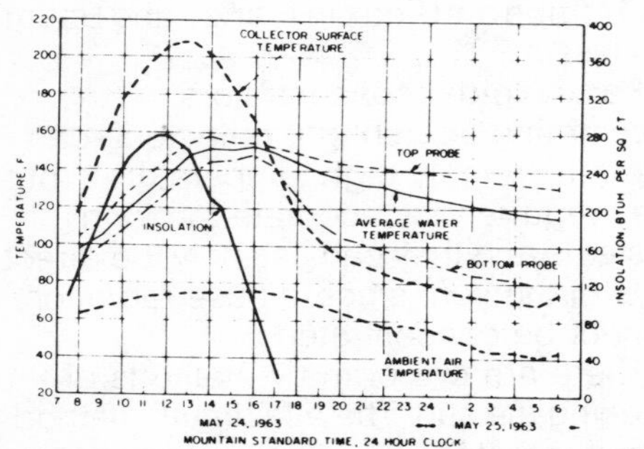


**Fig. 4 Performance of typical thermosyphon water heat at Phoenix, AZ on May 24 and 25, 1963. Forty gal. of water were heated to 150°F. (Courtesy HPAC)**

By the end of a typical sunny day, the storage tank is full of hot water at temperatures ranging from 120°F on a cold winter day to 165°F on a hot summer day. For single family residences, an absorber area of 16 to 20 sq ft and a storage tank capacity of 40 to 60 gallons have been found to give satisfactory results in Israel[22] and in Australia[14 23].

A number of precautions are necessary in the use of such a simple device as the thermosyphon water heater. Reverse flow downwards from the tank into the heater can occur on cold nights unless the tank is mounted at least 2 ft above the top of the collector[24]. Collectors with a single glass cover sheet are susceptible to freezing in cold climates and steps must be taken to prevent this. When air temperatures rarely fall far below 30°F, double glazing is generally adquate. For very cold climates, an indirect system must be used, with an antifreeze-water solution circulating through the heater and a coil-type heat exchanger in the domestic hot water tank.

Many years of experience in Israel[22] showed that desert dust could be a major problem, necessitating a tight seal between the cover glass and the collector casing. The combination of fine white dust and nocturnal dew can result in the formation of a hard, reflective white coating which can seriously reduce collector performance and must be removed mechanically. A thin layer of dust on the cover glass has very little effect upon the output of hot water.

The early versions of collectors used in the U.S. and Israel employed continuous tubes in sinusoidal patterns, which tend to heat a relatively small amount of water to a temperature that is high enough to cause deposition of minerals and clogging in the upper part of the collector. The grid system shown in Fig. 2 (B, C) has many advantages, including accessibility of the headers for cleaning, better coverage of the plate surface and relative insensitivity to clogging. Experience has shown that tubes smaller than ½ in. standard size should not be used. The day-long efficiency of a well-constructed and properly designed thermosyphon water heater will range from 45 to 65%, depending upon outdoor air temperature and wind velocity.

## FORCED CIRCULATION WATER HEATERS

There are many applications where the capacity of a thermosyphon system with one or two collector panels is inadequate to meet an existing demand for hot water. There are other situations where architectural or other considerations require that the storage tank be below or at a considerable distance from the collectors. In these cases, forced circulation systems are employed[23], with flow rates in the order of 1 gallon/hr per sq ft of absorber surface. Centrifugal hot-water accelerator pumps have proved to be satisfactory at these relatively low flow rates.

When a number of collector units are to be operated simultaneously, series-parallel arrangements have been found to be superior to operation with all of the collectors in parallel. Australian authorities with much experience in this field[23] report that up to 24 tubes may be operated in parallel; for larger numbers, downcomers may be used or multiple parallel system may be employed. Provision of vents to avoid air-binding is essential.

## FORCED-CIRCULATION AIR HEATERS

Solar air-heaters have not yet received the amount of detailed study that has been given to water heaters but there are many applications where air is a more suitable heat transport fluid than water. Air cannot freeze and air leakage is annoying but not as serious in its consequences as are water leaks. Natural circulation of sun-heated air is being employed in the French houses at Font Remeu, near the great solar furnace at Odeillo. Since the specific heat of air is only 0.24 Btu/1b/F as compared with water's 1.0 Btu/1b/F, and its density is only about 0.075 1b/ft[22] as compared with water's 62.4 lb/ft[22], it is obvious that the ducts needed to convey air to and from a solar collector must be far larger than the pipes used to carry water. However, most air conditioning systems rely upon air to do the actual heating and cooling, and so air has some notable advantages as a heat transport fluid.

Fig. 2 (K) shows a finned plate heater in which the air flow is generally parallel to the direction of the channels. In some cases, it is advantageous to allow the air to flow on both sides of the heat-absorbing plate and it may well be desirable to allow the air to enter at both ends of the horizontally-finned or corrugated heater and draw it out with a fan connected to a plenum at the center of the collector[25 26 27].

Fig. 2 (L) shows a matrix-type air heater in which the sun's radiation is absorbed by blackened gauze or other foraminous material; air is drawn through the system[25].

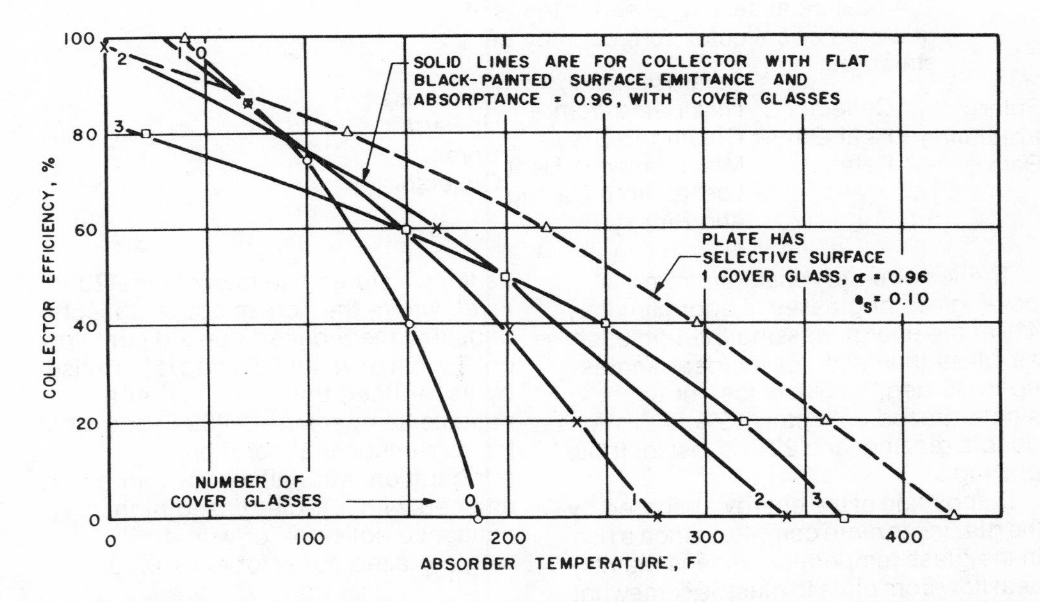


Fig. 5 Typical efficiencies for flat plate collectors under 300 Btuh/ft² insulation, with 70 air and sky temperatures. (Courtesy D. Erway and McGraw-Hill Book Co.)

Fig. 2 (M) shows the overlapping glass plate air heater devised by Löf,[1] in which the lower portion of the plates is darkened to absorb the solar radiatio which passes through the clear upper portion of the overlying glass. An alternative to this system uses overlapping blackened or selectively-surfaced aluminum sheets, with the exposed upper portion of the sheets acting as the absorbing surface and the covered lower portion serving as additional heat transfer area.

Corrugated steel or aluminum collectors, such as those shown in Fig. 1 (E and F) can also be used for air and they offer the advantage of relatively low first cost. The Australian scientists of the Commonwealth Scientific and Industrial Research Organization (CSIRO) have pioneered in the use of vee-corrugated air heaters using selective surfaces, as shown in Fig. 2 (N).

## COLLECTION EFFICIENCIES

The efficiency of a solar collector may be defined as the ratio of the amount of heat usefully collected to the total solar irradiation during the period under consideration. Instantaneous efficiencies during the middle of the day, when the incident angle is favorable, are generally higher than day-long efficiencies which must take into account the high and unfavorable incident angles which prevail during the early morning and late afternoon hours. Detailed mathematical analyses of collector efficiency calculations are to be found in[1 4 19]. A highly simplified approach to this very complex subject may be presented in the following terms:

| Collector Heat Gain, Btuh/ft² | = | Heat Transport Fluid Flow Rate, Lb/hr/sq ft | × | Specific Heat of Fluid | × | Temperature Rise of Fluid, deg F |
|---|---|---|---|---|---|---|

A heat balance, expressed in terms of unit area of collector surface, may be stated as:

| Solar Irradiation Rate | = | Collector Heat Gain Rate | + | Solar Radiation Dissipated from Glazing to Sky Miscellaneous Heat Losses from Casing and Piping | + | Heat Loss Upward from Collector |
|---|---|---|---|---|---|---|

(19)

The loss due to reflection from the cover glass or glasses is approximately 4% of the energy passing through each air-glass interface, for incident angles up to 35 deg, so 8% is lost for a single-glazed collector, 15% is lost for double glazing, and 22% is lost for triple glazing.

Disposition of the energy absorbed by the glazing is more complex, since a rise in the glass temperature means that the heat flow from plate to glass is somewhat reduced. When absorption is taken into account[1] the losses for incident angles up to nearly 40 deg are: for single glazing, 10%; for double glazing, 18%; for triple glazing, 25%. The use of thin plastic films would reduce these losses significantly, but such films have the disadvantage of transmitting from 30 to 40% of the longwave thermal radiation emitted from the collector plate. The combination of an outer glass plate and one or more inner plastic films has interesting possibilities[1 2].

The major component of the losses from a well-insulated collector is the upward heat flow from the plate to the atmosphere. The upward heat flow is a function of the emittance of the plate, es, for longwave radiation, the temperature difference between the plate and the air above the glazing, and the wind velocity.

The amount of radiant energy absorbed by the collector plate is the product of the incident irradiation, Its, the transmittance of the glazing, $t_g$, and the plate's absorbtance for solar radiation, $\alpha_s$. The absortance is usually well above 90% (see Table 1) for incident angles up to about 40 deg, but above that angle both the absorbtance and the transmittance drop off rapidly. The efficiency of collection for a conventional plate and tube heat absorber, with good edge and back insulation, may be estimated from Fig. 5.

For very low fluid temperatures, which may be below the temperature of the ambient air, the efficiency is highest without any cover glass.

As the absorber temperature rises, the efficiency drops off rapidly with a single cover glass and a non-selective absorber. The addiction of a good selective surface, $e_s = 0.10$, makes a very marked improvement in efficiency as temperatures rise towards the 200°F level, where the non-selective collector finds that the reduced upward heat flow coefficient of double glazing is just offset by its reduced transmittance. At a collector temperature of 250°F, desirable for conventional absorption refrigeration, 40% efficiency can be attained with a triple-glazed high emittance collector, or with a single-glazed collector $e_s = 0.10$.

## CONCENTRATING COLLECTORS

Temperatures far above those attainable by flat-plate collectors can be reached if a large amount of solar radiation is concentrated upon a relatively small collection area. The use of simple flat reflecting wings can increase the amount of direct radiation reaching a collecting surface to a marked extent, but, because of the apparent movement of the sun across the sky, concentrating collectors must be able to "follow the sun" during its daily motion.

Reflective troughs with parabolic cross-sections, running east and west, require the minimum amount of adjustment since they need only to be corrected for changes in the sun's declination. There is inevitably some degree of morning and afternoon shading of the ends of the collection tube which is generally mounted along the focal line of the parabolic trough, but satisfactory results have been reported by Trombe and Foex in France and Baum in Russia.

Paraboloidal concentrators, resembling searchlight reflectors, can attain extremely high temperatures, but they require very accurate tracking systems and they can use only the direct rays of the sun, since diffuse radiation cannot be concentrated.

There are two primary methods of tracking the sun. The altazimuth method requires that the tracking surface change both is altitude and its azimuth to follow the sun in its motion across the sky. In an equatorial mounting, the axis of the concentrator is pointed to the north and arranged to change its tilt angle to compensate for the varying declination of the sun. The daily motion is simply rotation at the rate of 15 deg per hour, to compensate for the sun's apparent motion at the same angular rate. Daniels, in Ref. 45, pp 57-86, gave a very useful discussion of the use of focus collectors for the production of moderate temperatures. The horizontal parabolic trough, oriented east and west, has good possibilities, particularly in the form shown by Farber[27] in which a multiplicity of small troughs are mounted in a fixed frame, inclined at the angle of the local latitude. It is reported that very little adjustment is needed if the absorbing pipe is relatively large. Farber also reports the use of an equatorially-mounted parabolic trough, which intercepts a 6 ft by 8 ft beam of solar radiation and heats fluid to about 600°F.

The principal use of concentrating collectors in the past has been in the production of steam or high temperature fluids for use in refrigeration or power generation. The higher cost and added mechanical complexity of collectors which must follow the sun, and their inability to function at all on cloudy or overcast days, are disadvantages which must be borne in mind.

## HEAT STORAGE SYSTEMS

For about 50% of the hours of the year, any given location is in darkness and so some means of heat or energy storage must be used if continuous operation is essential, as it is with a solar heating or cooling system. A thermosyphon hot water heater has its own built-in storage tank, but some separate means of heat storage must generally be provided by solar-operated thermal systems. There are two feasible systems for storing heat in the quantities needed for house heating and cooling.

## SPECIFIC HEAT STORAGE

The first and most widely-used system employs tanks of water or beds of rocks to store heat by virtue of a change in their temperature. The specific heat of water is 1.0 Btu/deg F/lb and its density is 62.4 lb/ft[22], so a single cubic foot of water can store 624 Btu if its temperature is raised by 10 deg F. A 1,500 gal storage tank contains somewhat more than 6 tons of water (12,518 lb) and a temperature rise of 50°F would give the tank a storage capacity of 625,900 Btu. Obviously, the tank must be well insulated to minimize the loss of heat, and a pump must be used to transfer the water from the tank to the collector; gravity will generally return the water after it has been heated.

Beds of rocks provide an alternative means of storing heat when air is used as the transfer fluid and fans are available to overcome the pressure drop required to force or draw the air through the bed. The specific heat of rocks and most other solid materials is close to 0.20 Btu/lb/F, and a cubic foot of solid rock, weighing 100 lb, has a heat capacity of 20 Btu per deg F rise in temperature. Daniels[7] noted that a pile of closely packed spheres leaves about one-third of the volume unoccupied as void space, and this fraction is relatively independent of the size if the spheres all have the same diameter.

One solid cubic foot of rock has only 6 sq ft of surface but the surface is greatly increased if the material is used in the form of 2-in. spheres. Löf[30] has made extensive studies of the use of rock beds for heat storage and, in his Denver House[1], he employed two cylinders of fiberboard (concrete-column forms were used), each 18 ft high, to contain 11 tons of 2 to 1½ in. gravel.

In Australia, extensive use[31] is being made of rock pile storage systems in connection with both solar heat collectors and RBR (rock bed regenerative) cooling systems. Choda and Read[32] have reported on the performance of a solar air heater with 20 absorbers, similar to Fig. 2 (N), each with a glazed area of 6 ft by 2.5 ft arranged in series to provide 300 sq ft of surface, to provide heat for a 340 cu ft rock storage system. The rocks, consisting of ¾ in. clean basalt screening, are contained in a galvanized iron tank, 9.75 ft in diameter, 6 ft high, lined internally on the sides and top with 1 in. polystyrene foam insulation. The rocks are supported on a ⅜ in. by 14 gauge wire mesh, carried by a number of bricks which form a plenum chamber beneath the rock bed. The density of the rocks in the bed is 89 lb/cu ft and their specific heat is 0.21 Btu/lb/F. With an 80°F temperature rise, the heat storage capacity of the system was estimated at 510,000 Btu.

Thomason[18] employs some 50 tons of river rock around a 1,600 gallon tank of water to provide both low cost heat transfer surface of large area and additional storage capacity for his SOLARIS solar-heated houses in Washington, D.C.

The two heat storage systems described above, used separately or in conjunction, utilize the ***sensible heat*** capacitiy which all materials possess, but its limitations are set by the fact that specific heats of usable substances range only from 0.20 to 1.0 Btu/lb/deg F. The ***heat of fusion*** which is involved in the freezing and thawing of water (144 Btu per lb, at 32°F) or certain sodium salt-hydrate systems (90 to 118 Btu/lb at 96 to 122°F) provides a method of storing far greater quantities of heat in a given weight and volume of material.

For refrigeration systems, water makes a useful "cold storage" material and an ingenious system for compensating for the well-known expansion characteristics of freezing water is described by Davis and Lipper.[33] Their application involved the use of a solar air heater in conjunction with an air-to-air heat pump. The low-temperature heat storage provided by their expandable cans of water minimized the power demand during the cooling cycle while their simple air heaters (type 2-F) with weatherable polyester film glazing provided significant improvement in both capacity and Coefficient of Performance during the heating cycle.

## HEAT OF FUSION STORAGE

For heating systems, the extensive studies performed by Dr. Marie Telkes[34] [35] give a clear understanding of the possibilities of virtually all available salt hydrates, beginning with the familiar "Glauber's salt" ($Na_2SO_4.10\,H_2O$) which melts and freezes at 91.4°F with a heat of fusion of 108 Btu/lb. This cheap and plentiful material ($35 per ton in carload lots) when mixed with 3 to 4% borax to serve as a nucleating agent, costs less than 2¢ per lb. of mixture, and its melting-freezing temperature is 89°F.

Other combinations of sodium salts and nucleating agents have melting points which cover a wide range of temperatures, making some suitable for "cold storage" on the low temperature side of an air conditioning sytem while others are useful for storing the heat gathered by solar collectors. The key to successful use of heat-of-fusion materials lies in understanding the manner in which they operate. Stratification is a common characteristic, in which the solid constituent separates from the remaining liquid. When water freezes, the ice is less dense than the liquid and so, fortunately, it rises as it solidifies (otherwise our lakes and rivers would freeze from the bottom up instead of from the top down!).

Dr. Telkes has found that, when Glauber's salt is used in large containers, it also stratifies, but with the solid phase at the bottom because its specific gravity is 2.66, while the remaining liquid phase is less than half as dense. This problem can be overcome by using tubular containers which are long and thin, enabling their exterior surfaces to serve as heat exchangers. The surface-to-volume ratio must be high, in the range of 25 sq ft of heat transfer area per cu ft of volume.

The heat storage systems in the University of Delaware house[36] use elongated plastic tubes, similar in size and appearance to fluorescent lamps, to contain the eutectic salt hydrate mixtures. "Cold" will be stored in one bank of tubes which will be chilled by air coming from a heat pump operating only at off-peak periods. Heat will be stored for winter nights in another bank of tubes containing a different hydrate which will be melted during the day by heat from the air heaters on the south walls and roof of the house.

## COOLING BY SOLAR ENERGY

The use of heat to produce cold has been known for well over a century. Abel Pifre is reported to have used some of the steam produced by his solar boiler in Paris in 1878 to operate a primitive absorption refrigerator and produce a small quantity of ice. Since his time, other workers have explored three processes by which the sun's radiant energy can produce cooling effect. All three systems are widely used with conventional energy sources.

The first is the steam jet system, investigated in 1936 by W. P. Green[37] at the University of Florida. A concentrating collector was used to produce steam at a pressure high enough to make a steam jet ejector function, and this in turn caused evaporation and chilling of water in a tank connected to the ejector. The Coefficient of Performance proved to be low and the requirement for a sun-following concentrator made the system impractical from a commercial point of view.

The second system employs the familiar compression refrigeration cycle, driven by a Rankine cycle engine or turbine, operated by steam or some other vapor which can be generated in a solar collector. Baum and Kirpichev, using a large sun-following paraboloidal concentrator erected in Tashkent, produced more than 500 lb of ice per day, but the apparatus was far too costly and complicated to cause others to follow their lead.

The availability of fluorinated hydrocarbons as working fluids has reawakened interest in solar-powered compression refrigeration. Teagen and Sargent[38] have proposed an ingenious system in which refrigerant-type working fluids, particularly R-114, will be used for the power cycle while R-22 will be used in the refrigeration cycle. In one version of their system, a four-cylinder reciprocating engine is used, with two cylinders providing the power while the other two compress the R-22. They envision a system which can also produce electric power in winter when cooling is not needed. With a solar collector temperature of 120°C (248°F) and an air-cooled condenser, they aspire to a Coefficient of Performance as high as 0.4, which, while it is low compared to motor-driven compression systems, is good in comparison with absorption systems.

A comprehensive review of solar-powered refrigeration has been published by R. K. Swartman[39] and this concentrates most of its emphasis upon the various kinds of absorption systems which have been tried during the past several decades.

Löf[1] discusses the operation of commercially available absorption refrigeration systems using water vapor as the refrigerant and lithium bromide solution as the absorbent. Tests conducted at the University of Wisconsin by J. A. Duffie showed that hot water from a solar collector at 175°F could produce 2 tons of refrigerating effect from a commercial unit rated at 3 tons when supplied with heat at 250°F. Despite the lowered rating caused by the reduced temperature, the Coef. of Performance remained at 0.6.

The University of Florida has concentrated the attention of its Solar Energy Laboratory upon the ammonia-water absorption cycle and Farber[40] reports the successful operation of a continuous refrigeration system in which the solar absorber is also the generator which drives the ammonia out of the water solution. The Florida systems provide cooling with lower collector temperatures than any of the other systems thus far reported and they appear to be well adapted to use with the type of flat-plate collectors which can be built today.

Intensive research is going to be needed to produce solar air-conditioning systems which will be low enough in cost and high enough in performance to compete with conventional systems, but solar cooling has the great advantage of having the largest supply of energy available when the demand is highest.

Dubin[41] in his discussion of solar heating and cooling of large office buildings, notes that absorption equipment is now available in the 100-ton range which can operate satisfactorily at water temperatures of 210 to 220°F, rather than the 270 water or 15 psig steam which is required by conventional units. The capacity of the units is somewhat reduced, however. He anticipates the development of absorption units which can operate successfully at 180 to 190°F. Farber reports that this has already been accomplished with units of a size suitable for single-family residences, where the roof area is generally more than adequate for the necessary collectors[42].

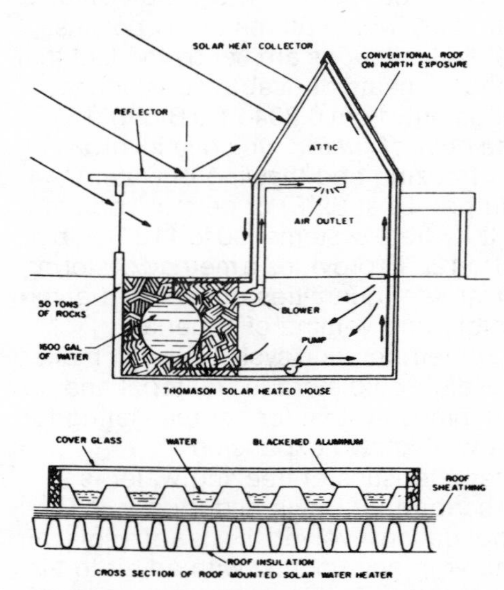


Fig. 6 Thomason solar heating system using roof-mounted single-glazed collectors with open water flow. Heat storage is provided by 1,600 gal water tank plus 50 tons of rocks. (Courtesy HPAC)

## COOLING BY NOCTURNAL RADIATION AND EVAPORATION

Compression refrigeration is the largest single user of electric power in air-conditioned homes, and many attempts have been made to use natural power-free phenomena to achieve comfort under summer conditions.

Yanigimachi[43] in Tokyo and Bliss in Tucson[44] (see also[1]) have used unglazed tube-in-strip collectors mounted on south-facing roofs to collect heat at moderate temperatures in winter and to reject heat in summer by radiation to the night sky. In both cases, radiant ceilings were used, with water tanks for storage and heat pumps to raise or lower the temperature of the circulating water as needed. The analysis of nocturnal radiation made by Bliss has already been mentioned[45] and this phenomenon can produce cooling at the rate of 20 to 30 Btuh/ft$^2$ during nights when the dew-point temperature is low.

During the summer of 1968, the combination of nocturnal radiation and evaporation was sufficiently effective to keep a prototype Skytherm building in Phoenix[16] within the comfort zone during 90% of the summer hours, and below 85°F at all times despite the fact that the afternoon temperatures consistently exceeded 110°F. No electric power was needed with this system, although the expenditure of 0.10 hp for the operation of a small fan-coil unit improved the senation of comfort by creating a moderate amount of air movement. The first full-scale Skytherm residence is now in operation at Atascadero, CA, as a joint project of the inventor, Harold Hay, Calif. State Poly. Univ. at San Luis Obispo, and HUD.

In Australia, CSIRO[46, 47] reports successful operation in 100 school cooling systems which use evaporation of water in the discharge air to chill the rocks in a switched-bed rock-filled recuperator. The incoming fresh air required for ventilation in Australian schools is generally considered to be 20 to 25 cfm per person, the higher figure being used for "infant and primary schools because of the higher body odours which occur". With the RBR system, 100% make-up air is used during the summer months and the cfm per pupil ranges from 28 to 85 cfm.

In the evaporative RBR (rock bed regenerator), the incoming air is cooled as it passes through rocks which have been chilled as the evaporatively-cooled exhaust air flows through the RBR. The air flow is switched every ten minutes. Water is admitted to the sprays for only ten seconds at the beginning of each cycle, and this is adequate to provide the necessary cooling. Power for the two fans used in the 2,000 cfm system is only 600 watts, since the pressure drop through the 5 in. deep pebble bed (¼ in. screenings) is only 0.13 in. The power consumption per sq ft of floor area is 0.75 to 1.0 watt, instead of 8.8 watts for mechanical refrigeration.

Thomason[17, 18] reports some success in cooling during nights with relatively low dewpoints by allowing water from his 1,600 gallon storage tank to run down the unglazed north slope of his roof. The combination of evaporation and nocturnal radiation produces some cooling effect, but Washington summers are characterized by too much humidity to promise much relief by natural cooling.

The conventional evaporative cooler was widely used in the southwest before the advent of motor-powered compression refrigeration. It is still used to cool industrial buildings and very low-cost housing but the high indoor humidities caused by direct evaporative cooling produce nearly as much discomfort as the 100-plus temperatures in which it operates.

## SOLAR HEATING AND COOLING SYSTEMS

It has been shown by Tybout and Löf[48] that the economic prospects of a combined solar heating and cooling system are much more favorable than for a system which performs either function alone. The cost of the collectors, from $2.00 to $4.00 per sq ft, can be justified much more readily when they are in use throughout the entire year than when they operate only during the winter months.

The simplest system which can accomplish both heating and cooling with the same equipment was used in the Hay Skytherm house which was tested in prototype form in Phoenix for eighteen months during 1967 and 1968[16]. This system is primarily suitable for use in the southwest where snow is not encountered and dewpoint temperatures are characteristically low. It uses a water-covered roof to collect solar radiation during the winter and to reject the daytime accumulation of summer heat by radiation and evaporation to the sky. Movable horizontal panels of weatherable urethane insulate are used to retain the heat collected on winter days and to prevent the sunshine of summer days from being absorbed by the water. During many weeks in the spring and fall, no movement of the insulation is needed because of the natural "thermal flywheel" effect of the heat storage in the building and its water-roof. The Skytherm system is capable of many modifications and the cooling effect of water which is subjected to both nocturnal radiation and evaporation must not be underestimated.

The Thomason system, Fig. 6, uses the entire south-facing area of a steeply pitched roof to support a very simple and inexpensive collector which is shown schematically in Fig. 2 (J). The collector surface is made of corrugated aluminum, lightly embossed to improve the uniformity of the water flow, and painted with an etching vinyl primer and a flat black paint. Since the collector is mounted directly on the roof sheathing, which is well insulated with glass wool between the rafters, no further insulation is needed. Many of the construction details are given in Ref. 18, along with a considerable amount of information gained during more than a dozen years of experience with houses of this type.

The system requires a pump to circulate the water from the 1,600 gal storage tank up to the ½ in. copper distributor pipe which has small holes to feed the proper amount of water to each trough. Gravity brings the warmed water back to a sediment-trap, from which it runs into the storage tank. The house itself is warmed and cooled by a forced circulation air system. The air receives its heat in winter from the 50 tons of egg-sized rocks which fill the bin surrounding the water tank. The rocks serve both as additional heat storage and as heat transfer surface.

Cooling is accomplished in the Thomason system by using a small conventional compression refrigeration unit which blows cold air through the rock bin, cooling the water in the tank in the process and chilling the rocks to the point where, during the following daylight hours, they can in turn cool and dehumidify the air in the house. The compressor motor is timed so that it does not operate during the afternoon peak hours.

In favorable locations, Thomason proposed to use nocturnal cooling of the water in the storage tank by allowing it to flow over the north-facing roof. Little or no cooling can be accomplished by operating the glazed south-facing collector at night, since the glass prevents both evaporation and radiation from the water to the night sky.

The three Thomason houses in a suburb of Washington, D.C. have accumulated far more operating hours than any other type of solar heating system. Their effectiveness has been demonstrated by the fact that the oil bills for their auxiliary heaters are far lower than those of their neighbors with conventional residences of comparable size. Data on performance of the individual components are needed, however, to enable the usefulness of the Solaris system to be evaluated for other locations.

The solar house at the University of Delaware, "Solar One"[36], uses cadmium sulfide cells mounted under twin glazing on its south-facing roof to generate electricity and to collect solar heat. Although CdS cells are recognized as being considerably less efficient than the single-crystal silicon cells which are used in the space program, it is believed that they can be mass-produced cheaply enough to make their use economically feasible in this application.

During winter days, the air flow needed to keep the cells at a safe operating temperature will be provided by a blower which will circulate the air through a eutectic salt with a melting point of 120°F. Six vertical south-facing panels on the south wall of the house will also provide additional winter heating.

During the summer, the house will be cooled by a heat pump operated at night by storage batteries which will be charged during the daylight hours by the solar-electric panels on the roof. Another eutectic salt, freezing at about 50°F, will be used for "cold" storage.

This house, put into operation during the summer of 1973, represents the first attempt to generate substantial amounts of electricity with photovaltaic cells which also produce heat from the excess solar radiation. It has been designed to obtain 80% of its thermal and electrical requirements from the sun; the balance will come, primarily at off-peak periods, from the local electric utility.

When both winter heating in severe climates and summer refrigerated air conditioning are required, a system similar to that shown in Fig. 7 may be needed. A high performance collector, preferably double-glazed and selectively surfaced, will keep a hot water or other storage system charged with heat at the highest attainable temperature. The domestic hot water supply may be obtained from a coil mounted in this tank or from a separate thermosyphon heater.

Water or air from the storage system is circulated by a pump or fan through an auxiliary heater to a valve or damper which will direct the heat transfer fluid through the heating system in winter or through the absorption air cooling unit in summer. A condenser, not shown in Fig. 7, is an essential part of the absorption refrigeration system. For large units, a cooling tower will probably be used to provide the condenser with water cooled nearly to the wet bulb temperature. Air cooled condensers must operate at temperatures above the dry bulb temperature, with the result that the refrigerating capacity and the Coef. of Performance are reduced.

The absorption system may use lithium bromide, ammonia and water, or other combinations of absorbants. The principle requirement is ability to function with a reasonably high Coefficient of Performance with a heat transfer flud temperature at or below 200°F.

Appropriate control devices will be needed to turn on the heater circulating pump only after the sun has warmed the collector sufficiently, and to turn it off when there is insufficient solar radiation. Conventional thermostats will be needed to control the air heating and cooling units. The auxiliary heater is more likely to be energized by electricity than by a fossil fuel, since it is highly probable that the use of natural gas will be curtailed rather than expanded in the future and light of oil of low sulfur content is also going to be short in supply and high in cost.

A solar heated and cooled structure, like an "all-electric house," will require as much insulation as can be put into it, to minimize heat loss in winter and heat

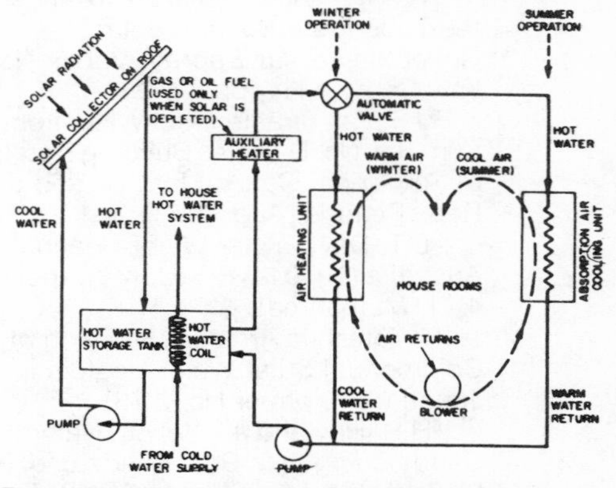

Fig. 7 Residential heating and cooling with solar energy: Schematic diagram of one alternative. (Courtesy NSF/NASA and the University of Maryland)

gain in summer. Protection from excessive solar heat gains will be absolutely essential and this will mean minimal use of glass on east and west sides. Insulating glass on the south side can be beneficial if it is shaded from the sun's direct rays in summer.

## CONCLUSIONS

A map of solar distribution throughout the U.S. shows that the annual mean daily insolation on a sq ft of horizontal surface ranges from a high of 500 Langleys/day (1850 Btu/day) in the Southwest to a low of 250Ly/day (925 Btu/day) in the rainy Pacific northwest. Throughout most of the ation, the average daily irradiation is in excess of 350 Ly/day (1290 Btu/ft$^2$/day). For a residence with 1,000 sq ft of roof area, the average annual insolation is considerably more than 1,000,000 Btu/day, which is far more than the average heating requirements.

Up to the present time, the use of solar energy has not been attractive because of the high cost of the necessary collection equipment and the necessity of providing 100% standby capacity for use during protracted periods of bad weather. Conventional fuels have been so plentiful and inexpensive that there has been little incentive to conserve them or to consider the use of solar radiation. That situation is changing rapidly, and it is to be hoped that the foregoing summary of solar energy utilization will be helpful to the many engineers who are going to be involved in this field in the future.

### REFERENCES


[1]A.M. Zarem and D.D. Erway; ***Introduction to the Utilization of Solar Energy*** (McGraw-Hill, New York, 1963).

[2]F.E. Edlin; Plastic Glazings for Solar Energy Absorption Collector (***Solar Energy***, Vol. 2, No. 2, 1958, P. 3).

[3]G.P. Mitalas and D.G. Stephenson; Absorption and Transmission of Thermal Radiation by Single and Double Glazed Windows (Res. Paper 173, Div. of Building Res., Nat. Res. Council of Canada, Ottawa, 1962).

[4]***Handbook of Air Conditioning Fundamentals;*** ASHRAE, 1972, New York.

[5]H. Tabor; Selective Radiation I. Wavelength Discrimination (Trans. Conf. of Sci. Uses of Solar En., 1955, Vol. 2, Part 1-A, pp. 1-23, Univ. of Ariz. Press, 1958).

[6]J.T. Gier and R.V. Dunkle; Selective Spectral Characteristics as an Important Factor in the Efficiency of Solar Energy Collectors (ibid, pp. 41-56).

[7]F. Daniels; ***Direct Use of the Sun's Energy*** (Yale Univ. Press, 1964).

[8]D.K. Edwards, et al; Spectral and Directional Thermal Radiation Characteristics of Selective Surfaces (***Solar Energy,*** Vol. VI, Jan.-Mar., 1962, No. 1, pp. 1-22).

[9]E.A. Christie; Spectrally Selective Blacks for Solar Energy Collectors (Paper No. 7/81, 1970 I.S.E.S. Conf., Melbourne).

[10]C.P. Butler, et al; Surfaces for Solar Spacecraft Power (***Solar Energy***, Vol. 8, No. 1, Jan.-Mar., 1964, pp. 2-8).

[11]H.C. Hottel and B.B. Woertz; The Performance of Flat-Plate Solar Collectors (Trans. ASME, Feb. 1942, Vol. 64, pp. 91-105).

[12]A. Whillier; Thermal Resistance of the Tube-Plate Bond in Solar Heat Collectors (***Solar Energy***, Vol. 8, No. 3, July-Sept., 1964, pp. 95-98).

[13]J.F. Van Straaten; Hot Water from the Sun (Ref. No. D-9, Nat. Building Res. Inst. of So. Africa, Council for Ind. and Sci. Res., Pretoria, Aug. 1961).

[14]E.T. Davy; Solar Water Heating in Australia (1970 I.S.E.S. Conf., paper No. 4/71, Melbourne).

[15]I. Tanishita; Present Situation of Commercial Solar Water Heaters in Japan (ibid, paper No. 2/73).

[16]H.R. Hay and J.I. Yellott; Natural Air Conditioning with Roof Ponds and Movable Insulation (Trans. ASHRAE, Vol. 75, Part I, 1969, pp. 165-177).

[17]H.E. Thomason; Three Solar Houses (ASME paper 65-WA/SOL-3; see also ***Solar Energy***, Vol. IV, No. 4, Oct. 1960, pp. 11-19 and Ref. 6, p. 279).

[18]H.L. and H.J.L. Thomason; ***Solar House Plans*** (Edmund Scientific Co., Barrington, N.J., 1972); see also, by the same authors, Solar Houses/Heating and Cooling Progress Report (***Solar Energy***, Vol. 15, No. 1, May 1973, pp. 27-40).

[19]K.G.T. Hollands; Directional Selectivity, Emittance and Absorbing Properties of Vee Corrugated Specular Surfaces (***Solar Energy***, Vol. 7, No. 3, July-Sept., 1963, pp. 108-116).

[20]G. Francia; A New Collector of Solar Radiant Energy (U.N. Conf. on New Sources of En., Rome, Vol. 4, p. 572, 1961).

[21]Buchberg et al; Performance Characteristics of Rectangular Honeycomb Solar Thermal Converters (***Solar Energy***, Vol. 13, No. 2, May 1971, pp. 193-217).

[22]J.I. Yellott and R. Sobotka; An Investigation of Solar Water Heater Performance (Trans. ASHRAE, Vol. 70, 1964, pp. 425-433).

[23]R.V. Dunkle and E.T. Davey; Flow Distribution in Solar Absorber Banks (ibid, paper No. 4/35).

[24]D.W.N. Chinnery; Solar Water Heating in So. Africa (Bulletin No. 44, N.B.R.I., C.S.I.R., Pretoria, 1967).

[25]Maria Telkes; A Review of Solar House Heating (Heating and Vent., Sept. 1949, pp. 68-74).

[26]M. Telkes and E. Raymond; Storing Heat in Chemical: A Report on the Dover House (Heating Vent., Nov. 1949, p. 80).

[27]E.A. Farber; Solar Energy Conversion and Utilization (***Building Systems Design***, June 1972).

[28]R.W. Bliss; An Experimental System Using Solar Energy and Night Radiation (U.N. Conf., 1961, paper No. E-35-S-30; see also Ref. 6, pp. 276 and 282).

[29]R.L. Bliss; The Derivation of Several Plate Efficiency Factors (***Solar Energy***, Vol. 3, No. 4, Dec. 1959).

[30]G.O.G. Löf; Unsteady State Heat Transfer Between Air and Loose Solids (***Ind. Eng. Chem.***, Vol. 40, 1948, pp. 1061-1070).

[31]D.J. Close, R.V. Dunkle and K.A. Robison; Design and Performance of a Thermal Storage Air Conditioning System (Mec. and Chem. Trans., Inst. of Eng., Australia, Vol. MC-4, No. 1, May 1968, pp. 45-54).

[32]A. Chada and W.R.W. Read; The Performance of a Solar Air Heater and Rockpile Thermal Storage System (Paper No. 4/48, I.S.E.S. Conf., Melbourne, 1970).

[33]C.P. Davis and R.L. Lipper; Sun Energy Assistance for Air-Type Heat Pumps (Trans. ASHRAE, Vol. 64, 1958, pp. 97-110).

[34]Maria Telkes; Solar Heat Storage (Paper No. 64-WA/SOL-9, ASME, New York, 1964).

[35]M. Telkes; Storing Solar Heat in Chemicals (***Heat. and Vent.,*** Nov., 1949, pp. 80-86).

[36]K. Boer; A Combined Solar Thermal Electrical House (Int. Sol. Energy Congress, 1973, paper No. EH-108).

[37]W.P. Green; Utilization of Solar Energy for Air Conditioning and Refrigeration in Florida; Master's thesis, College of Eng., Univ. of Florida, Gainesville, 1936).

[38]W.P. Teagan and S.L. Sargent; A Solar-Powered Combined Heating and Cooling System (Paper No. EH-94, 1973 I.S.E.S. Congress, Paris).

[39]R.K. Swartman, Vinh Ha and A.J. Newton; Review of Solar Powered Refrigeration (Paper No. 73-WA-SOL-6, ASME, 1974).

[40]E.A. Farber; Design and Performance of a Compact Solar Refrigeration System (Eng. Progress at U. of Fla., Vol. XXIV, No. 2, p. 70, 1970, Gainesville).

[41]F.S. Dubin; Testimony before Subcom. on Energy, House Science and Astronautics Comm., June 12, 1973 (published by Dubin-Mindel-Bloom Associates, New York).

[42]E.A. Farber; The Direct Use of Solar Energy to Operate Refrigeration and Air Conditioning Systems (Tech. Progress Report No. 15, Florida Eng. and Exp. Station, Univ. of Florida, Gainesville, Nov. 1965).

[43]J.I. Yellott; Japan Applies Solar Energy (***Mechanical Engineering***, March, 1969, pp. 46-51).

[44]R.W. Bliss and Mary D. Bliss; Performance of an Experimental System Using Solar Energy for Heating and Night Radiation for Cooling (U.N. Conf. on New Sources of Energy, Rome, 1961).

[45]R.W. Bliss; Atmospheric Radiation Near the Surface of the Earth (***Solar Energy***, Vol. 5, No. 3, 1961, pp. 103-120).

[46]F.C. Hogg; A Switched-Bed Regenerative Cooling System (Proc. XIIIth Int. Conf. on Refrig., Vol. 4, Washington, 1971).

[47]W.E. Read, et al; Use of RBR Systems in South Australian Schools (Aus. Refrig., Air Cond. and Heating, Vol. 26, No. 12, Dec. 1972, pp. 20-27).

[48]R.A. Tybout and G.O.G. Löf; Solar House Heating (Nat. Resources Journal, Vol. 10, No. 2, April, 1970, pp. 283-326).

# Solar Energy - A Better Understanding

by Harold R. Hay

## Introduction

An unsubduable resurgence of interest in solar energy is encouraging evidence that the industrial revolution may soon attain maturity. In a brief span of time, precocious man has moved from control of fire to a rate of industrialization that threatens to exhaust all fossil fuels stored on earth since the world began. Had the present scarcity not raised prices for that former solar energy, stored as gas, oil, and coal, some more calamitous effect would have brought the profligacy to a halt — a major climate change, or an unbreathable atmosphere. We are fortunate that a gentle economic pinch awakened us.

To understand why we dozed —while racing toward environmental disaster-the precepts of the industrial revolution and the perversion of language that fostered it must be examined. The doomsday prophets were too nearly right for us now to do less; corrective measures, including solar energy use, cannot be effective unless we understand well the reasons for their need.

At incredible speed, the industrialized countries accepted changes that provided **power**—in both its political and energy contexts. The repercussions of the present realignment of power control, from the populated developed countries to the almost unpopulated developing ones, are yet to be realized. These repercussions may include life-style denial for the industrialized countries and life-denial for the over-populated, under-developed countries, such as India. The present energy crisis as well as its future consequences is still in the hands of companies and officials who, in recent years, have proved that they had no foresight to anticipate the chaos their judgements caused. They promised us availability of power, which we accepted without questioning, as if it were offered omnisciently.

Primitive man, through his traditional wisdom, was better prepared to cope with the complexities of energy effects in nature than are the ritualistic over-specialized energy experts of today. The moderns computerize the energy balance without feeling it; they attempt to surpass the efficiency of nature without awareness of implications. This idea-ritualism approaches a cult of high technology worship that denies common sense. The experts unquestioningly accept cheap power as a path to an ever better world, but they deny the virtue in methods of solar energy use that have operated the world outside man's economy for billions of years. They spurn low technology to justify their over-zealous specialization and their promotion of the high technology which has produced our problems—as well as our benefits. They now propose to use solar energy to operate inefficiently the non-natural devices which they designed to consume concentrated forms of energy.

Developing primitive man accepted the sun as part of his religion and, in Egypt, made it the basis of his first monotheism. Along with its lifegiving attributes, the sun was a symbol of power, and, as such, was overthrown by human gods which now appear to have yielded to new forms of power—forms directly available to us in our houses and cars, in our national policies and in our computers. Because we have had no long experience with handling such power, our infatuation with it is ill-advised and hazardous.

Those who exploit the world's fossil fuels have used alluring words to sell their ability to deliver as many "energy slaves" as we could pay for. The "all-electric" house was called a "Gold Medallion Home"—but we have gone off that gold standard also. Two to three times as much fossil fuel is consumed to generate and transit the electricity for thermal control in residences as is required for the same effect if the fuel were burned directly in the house. Unnecessary electric heating (and cooling) is approaching the status of a social evil; it is also becoming too costly for any except the well-to-do who presently have this thermal waste subsidized. No longer are we promised all the cheap energy we can hope to use. The words which built our hopes, and shaped our houses, had no established scientific substance within our time range—which is less than 200 years for natural gas and about 100 years for coal at reasonable production cost.

Research is respected most when applied to in-depth studies in fields of natural and physical sciences for determining verifiable truth about fundamental natural laws. The word, research, has fallen into disrepute through misuse of it to describe activities carried out by grade-school students, diletantes, and poll-takers—among others. Until recently, electric utilities were a delivery service not required by utility commissions to have a qualified research staff. Nor did the oil companies bother much with scientific research until they began competing for the petrochemical business. With the advent of the energy crisis, unqualified persons were quickly slid into organization gaps labeled "research department." Research is a highly specialized field; too often, in companies and sometimes in academic institutions it may be reviewed by higher executives unfamiliar with research ethics and interested primarily in window-dressing. In the energy field, these are the same type of executives whose short-sightedness did not encompass the present crisis. Few of them will admit publicly their culpability; their lack of training; their tunnel vision. The research director should be a person of **proved** vision—a rare bird who is often turned away because he intuitively foresees difficulties as well as benefits. Research today has too frequently become a game for sycophants and seekers of government contracts that provide sinecures.

The oil and utility officials, through ignorance of facts and through a fallacious confidence in the capability of subordinates to out-power nature, modified our attitudes to natural laws affecting thermal comfort and energy use. Unwittingly, most solar scientists contributed toward the downgrading of solar energy. They predominantly visualize the future of solar energy as being adjunctive to "conventional" heating and cooling equipment and as conformative to textbook concepts that have displaced man's ancient and tested knowledge of how to use the solar energy that reaches earth. As a result, the expert's attempt to add the complex to the over-complicated put solar energy at a double disadvantage—of being capital intensive (when it hasn't been before) and of being used for purposes for which it is not well suited. Consequently, these solar experts have produced too little of practical, present value for our energy situation. They have, thereby, erroneously convinced the energy suppliers and administrators that solar energy has little immediate potential for reducing our fossil fuel demand. With such advocates as solar energy friends, those who would be competitively hurt by its development have safely remained relatively silent.

There is justification for confining this discussion of solar energy to the heating and cooling of buildings. Two deterrents affecting the immediate use of solar energy are the confusion in the public mind concerning what is practical and a surfeit of publicity lauding the impractical. The heating and cooling of buildings with natural, diurnal energy forces is practical; the use of silicon cells or other photovoltaic conversion of solar radiation to electricity is impractical for heating or cooling of residences today and into the future because the conversion to electricity will never compete economically with direct use of solar heating and nocturnal cooling. Conversion is a wasteful process compared with direct use. For the next 10 or 20 years, the public should hear less about solar power; only research people and some special users who can afford exhorbitant prices for solar power will benefit within that time frame. Windmills come closer to being practical for producing the electricity needed for household appliances which cannot have their functions satisfied by direct use of solar energy.

Solar water heating has been in use for many years; there is, unfortunately, little new technology that warrants discussing more than the features and problems common with the use of flatplate collectors for space heating. Solar distillation is a practical application of solar energy to get potable water from the sea or from brackish sources. Solar stills have been abundantly discussed in the literature but public demand remains so low that a review of the subject would detract from the main potentials of solar energy today—the provision of thermal comfort in buildings.

**The Balance of Natural Forces**

So long as we think only in terms of using solar energy—including the minor potentials of wind, ocean-thermal, bio-thermal, or other—we shall be diverted from the best support for its use and we shall be following unthinking ritualistic patterns of primitive and mechanized man. This is best revealed by the monstrous folly of widescale air conditioning which wastes vast quantities of fossil fuels and scarce metals and has also deteriorated to an unrecognized degree the quality of life within buildings.

Few persons are grateful for the intermittent or continuous noise and strong drafts of central, or window-mounted, refrigeration cooling units. Nor do the assaulted ones get maximum comfort from having one side of their body subjected to colder air than the other side. The drier air produced by such refrigeration devices contributes to comfort during a humid period, but to discomfort during a dry spell. To reduce the cost of electric cooling, we have increasingly built ourselves into a cocoon of insulation which can have some undesirable consequences that will be discussed later. Because there are no heat-storage materials in the "cocoon," our modern rooms are not necessarily so comfortable as are those of stone castles, old Spanish missions, or the solid brick or concrete houses which preceded the modern structures based on flimsy, insulated panel walls having low first cost.

Certainly, a poorly-built or light-frame house can be made more comfortable by an air conditioner during hot summer months—but a well-built house with heavy wall construction will unquestionably have the potential for far greater comfort, in any part of the United States. To comprehend this, we must understand the properties of materials as well as the character of the daily energy flux.

We understand solar energy—to a degree. We enjoy it on a cool day and avoid it on a hot one. We more or less know that we are a solar collector and form vitamin D while getting a good tan; and we have felt the pain and feared the skin cancer that can be caused by excessive ultra-violet collection. We have sensed the high heat storage of black pavement, in July, and have seen paper burned by solar radiation focused by a magnifying glass. These observations indicate, in part, the solar energy with which we live and which we can use.

We also know that clouds may hide the sun for several days and that solar radiation is much weaker in the morning and evening than it is around noon. For these reasons, engineers turned away from solar energy as too unreliable, too low level. It is not continuous nor it is easily stored in a form useful to run conventional mechanical devices. Thank goodness, and thank nature. Solar energy is, however, easily stored in non-mechanical ways. If the sun were of a much higher level of intensity, if it were continuous day and night, there would be no life on earth.

Our indoor patterns, awake or asleep, prevent us from being actually aware of the nocturnal forces of nature—forces as important to life on earth as solar energy itself. Nocturnal cooling results from one or a combination of three phenomena which can be related to the black pavement which was so hot in the afternoon and so cold in the morning. It was not cooled by an air conditioner but by natural forces; ones that we can use to cool our buildings without gas, oil, electricity or complicated equipment.

The black pavement may be cooled by a breeze, just as one can cool a spoonful of soup by blowing on it. This convection cooling is especially powerful where sea breezes have low nighttime temperatures or where cold mountain air slides into valleys after sunset. Convection cooling can be used wherever summer nights are cool. Probably the earliest forms of animal life leaving the sea had to find protection from the sun during the day and from convection heat loss at night by hiding behind stones which formed the first land shelter—walls and roof.

The black pavement might also be cooled on a breezeless night by radiation to the sky. In arid regions, this is a powerful phenomenon that most people do not understand. The radiant heat waves leaving the pavement pass through the overlying air and out to cold space without heating the dry air above the pavement. Meteorologists are well aware of the frequency of frost on the ground when air temperatures are above freezing. In Iran, ice is made by radiation in open water pans (protected by walls from the sun and winds). At times the thin layer of water may freeze when overlying air is as warm as 47°F. Convection would not cause water to cool below 47°.

Water evaporation is a third cooling factor to be considered. The black pavement is rapidly cooled when rain removes heat as the liquid water changes to vapor which may be seen as steam rising into the air. This is similar to the cooling effect of conversion of water to steam in the tea kettle; an effect that carries away the heat added to the water by the stove. Evaporation cooling is directly sensed by a sweating person who fans himself. All water that has been evaporated from the sea, lakes, rivers, ground, etc., or from plants and animals, has provided a natural cooling effect that is a very significant protective factor in the balance of life. Cooling can also take place by conduction, an example is loss of heat from the hand in contact with a cold surface, such as a glass of ice water; conduction occurs as the rain washes across the black pavement but, eventually, this transferred heat may be lost in the convection, radiation, or evaporation processes.

Only a few decades ago, in the United States, and still the situation for most people of the world, these cooling methods were the only ones used by man. During the day they helped offset an excess of heat from solar radiation and their comfort effects are most apparent in shade, but day cooling is not so discernable as that at night. It used to be that we sat on porches in rocking chairs or swings to stir up currents of evening air that would cool us by convection and evaporation. We never have understood radiation cooling so well as animals do, though we may be aware of it when we are near a shaded stone or concrete building. The animal, however, may go further by also orienting its body to the "cold spot in the sky" which man has studied little and has not knowingly used for his own comfort. There may be other vitally important phenomena applicable to man's comfort, if we returned to the scientific observation of how life adapts to the climate. Such studies should become a vital part of our effort to conserve fossil fuels.

For millions of years, man successfully lived with only solar heat and nocturnal cooling to provide thermal comfort in some rather severe climates. During this time, he developed house designs which properly related building materials, and architectural form to the climate in order to obtain a high standard of comfort. The most thermally stable environment ever created by man is at the centers of the great pyramids of Egypt, where the temperature never changes throughout the year. Very large differences between day and night temperatures (over 100°F) exist on the pyramid's outside stone surfaces, with smaller differences deeper in the stone. Cyclic heat gain and loss on a diurnal and seasonal basis occur in the outer portion of the pyramid with a time lag resulting from the slow conduction in and out of dense materials such as stone.

The lower the density of the material, the slower is the heat conduction through it. Thus, foamed concrete, containing many air cells through which heat is poorly conducted, will not get so hot in depth as will dense concrete. The surface of the foamed concrete may be hotter than that of the dense since it receives the same amount of heat, but this excess heat reradiates or is carried away from the surface by convection cooling. The thermal balance is a result of many factors. Wood and thatch are low density and good insulators when dry but they are not so good as hair or feathers, nor so good as fiberglass, foamed plastic, or a vacuum. The last is too difficult to maintain for thermal control in buildings. These relationships of building materials to the natural energy flux provided excellent thermal conditions for life before man learned to control fire, to produce power, or to make refrigeration coolers.

### "Modern Times" Housing

When man began adjusting his thermal comfort to standards higher than those provided locally by nature, he began creating major problems. The heating of a few cold and damp caves did not create more than a self-limiting pollution, but the burning of fossil fuels for heating buildings in temperate and arctic regions and for cooling buildings in the tropics, should now be recognized as one of his accomplishments that is messing up the world. Our energy problem would be less serious had the industrial revolution occurred several thousand years earlier while civilization was centered in the mild Mediterranean climate area. Then, the population centers would have remained in a region of low energy demand for thermal comfort. Perhaps, though, man had to be uncomfortable in the northern climates before he was spurred to change. A danger today is that we may be temporarily so comfortable through wasting fossil fuels that we will not endeavor to further improve our comfort by means less wasteful. An inevitable problem man must soon face is whether he can maintain present occupancy of lands with climates to which he is not genetically adapted and which require large amounts of energy to obtain thermal comfort.

A second contributing energy crisis factor is substitution of horsepower for manpower. This core of our pride provides industrialized nations with a higher standard of living than that which the rest of the world's population can ever hope to attain. Twenty years ago, the writer pleaded with the Government of India not to adopt the building standards of industrialized countries, but to emphasize instead low-energy technology. This advice was not welcomed even though, as one example, it was shown that to attain the same per capita consumption of cement as industrialized countries, India would have to build more cement factories than exist in the whole world. With the present oil shortage, which may devastate the country, India will have to move strongly and rapidly in the direction of energy-conserving technology — as will, of course, industrialized countries.

In India, thermal comfort deterioration resulting from modernizing building practices was conspicuous. Metal or asbestos-cement roofs reduced fire hazards and the need for re-thatching; the health effects were mixed. Such modern roofs proved excessively hot in summer and cold in winter. To remedy these shortcomings, it was necessary to add insulation, the production and transport of which required high capital investment and high energy consumption. Low-cost, adobe-type walls were being replaced by costly and thin burned-brick walls; again, the occupant had lower comfort since neither fuel nor electricity were available to complete the modernization by re-establishing the thermal balance earlier provided with traditional construction.

Throughout the world (and the United States), certain of the oldest buildings should be protected by law; not just as historical monuments, but also as examples of indigenous architecture in which materials were used in optimum relationship with the climate. Our poverty of foresight in attempting to develop omniclimatic thermal control systems that foster use of wrong building materials should be recognized as a major cause of fossil fuel exhaustion. In the Southwest, indigenous, flat-roof architecture is ignored by most people moving there from the East or Midwest. For purely psychological comfort, they pay for a pitch-roof similar to that "back home" — mentally secure from some heavy snowfall that will never occur. They also unnecessarily pay to use heaters and coolers, sometimes both on the same day, because they didn't want the heavy (and soundproof) walls of indigenous-style architecture.

Heavy wall construction of stone, brick, and solid masonry is characterized by high heat storage and slow heat transfer; such construction is said to have "capacity insulation." When air gaps are left between wall wythes of external and internal masonry, a lighter weight wall and lower thermal transfer are obtained. This "cavity insulation" is further improved by filling the air space with loose insulation, such as vermiculite or perlite. The combination provides effective cavity insulation as well as the heat-storage capability of solid masonry on the room side of the wall where heat storage has great value in modulating temperature changes. Cost of manufacture, transport, and site labor, plus the marketing of thin sandwich walls for skyscrapers and frame construction for residences, produced a trend away from heavy walls. We now have what was termed earlier as "cocoon" buildings.

The cocoon, undeniably, is cheap in first cost and thermally comfortable, but only so long as a heater or an air conditioner is running. These appliances must operate almost continuously when there is little heat storage in the room. If a door is left open to a greatly different outside temperature, the room quickly attains the discomfort level of the outside, and there is an appreciable time lapse before comfort is re-established. A more serious problem exists when a blackout stops the input of electricity — even for a short time. Then, heaters and coolers stand idle while indoor temperatures reach those of the outside, or worse.

Imagine sunshine pouring into the cocoon through a south window (no windows are tolerated in the cocoons of nature) on a hot summer afternoon during a blackout. Due to the greenhouse effect, indoor temperatures would rapidly soar to 140°F or more; then occupants would need to open windows or such to cool off in the 90 or 110° outside temperature. (Your friendly utility man is unlikely to guarantee that blackouts will not be an increasingly frequent occurrence.) We have not yet been punished by this severe shortcoming of modern insulated construction — and of thin dome structures — which depart so strongly from man's traditional knowledge of how he can live in harmony with natural energy forces.

Another disadvantage of the cocoon is its dependency upon power during peak load hours when use of electricity will be penalized by special, very high rates. Some foreign countries already use this penalizing rate structure — one that is now being introduced into the United States by legislation. The rate increases are intended to be so high that afternoon use of electricity will be greatly curtailed. No politician enjoys enacting legislation that will so directly affect a great many voters. Hence, we do not have such rates in effect now and hear too little about their coming — a crisis is being awaited to make the bitter medicine acceptable.

We must, however, credit the politicians for having sought other alternatives and for moving faster than the majority of the public is inclined to go.

Zealously "modernizing," Americans became partial to window-walls: the over-size glass surfaces which heat or cool so much of the out-of-doors. Rising utility bills focus attention on these panes through which the homeowner seldom looks except for ten-minute intervals when showing a new guest around. Future prospective buyers will be less enthralled with energy-wasting houses, the value of which will depreciate abruptly. They will expect only a comfortable and needed amount of glass, as is found in more traditional buildings. Reflective glass is being accepted as a modern means to reduce undesirable summer heat gain (and also desirable winter heat gain). This reflects in law suits filed by owners of adjacent buildings whose air conditioners become inadequate to handle the "solar garbage" from the reflective glazing.

"Sun rights" are also headed for the courts. Many states and Congress are about to assure an owner of his full access to all sunshine potentially available to his property on the winter solstice of December 21st. Neighbors would not be permitted to reduce it by interception with buildings or trees on their own land. Property values will be increased or decreased by such legislation. The writer recommends that property deeds be required to specify the extent to which solar energy is available on property. A sun-drenched plot in the North will assuredly out-value a heavily shaded one. To know this when buying property may be as ethically important as knowing whether the property is under water or consists of sand dunes.

During the past forty years, in the United States, there has been a strong trend away from on-site labor and away from heavy building materials, such as brick, stone, and concrete. The modern panel construction makes possible lower first cost and it gets more houses built with less skilled labor; mortgage down payments and monthly charges are also less. But operating costs for poorer thermal comfort, for maintenance and depreciation, and for the energy which drains national resources are greater. Some persons advocated twenty years ago that a house should not last forty years because it would then be obsolete and should be replaced. An "economy of waste" was the source of momentary pride; we moved forward by bulldozing the old without recycling any part; thus, in part, we built the modern energy crunch.

Today, the more foresighted architects talk of building for use throughout one or two hundred years during which time there will be flexibility in the design for important but not supercilious change. These architects believe that modern buildings have made obvious most of the errors man should guard against. They are concerned not only with reduction of operating energy on a life-span basis, but also with eliminating, to the extent logical, building materials whose production has a high energy demand or threatens to exhaust scarce resources, including metals such as copper.

Architects are at a great disadvantage in these efforts because the public is not yet ready to accept new realities which cause use of solar energy and nocturnal cooling to be different from conventional heating and cooling. The public would like to have solar energy operate present heating and cooling devices so that limitations and advantages of local climates may be ignored. The public gave no support to those few who, over the years, advocated solar energy. The acceptance of solar heating and nocturnal cooling will now be slowed by this neglect. Time is required for architects to creatively demonstrate the hundreds of models of naturally air-conditioned houses from which the public may choose styles that help to make attractive low-energy communities.

### Flatplate Solar Collectors for Water and Space Heating

Young enthusiasts who consider solar energy utilization to be simple enough for quick benefits from do-it-yourself efforts are learning many lessons. Progress is being made along these lines; but know-how is not standardized, many wrong ideas are accepted, and many over-simplifications result in frustration. If there were no problems, obviously solar energy would be in wide use throughout industrialized countries. For over forty years, the basic design of flatplate collectors has been known: water circulation through metal tubing attached to a blackened metal sheet with an air space under a glass cover and a backing of insulation assembled in a frame. Tens of thousands of such collectors were used in California and Florida; later hundreds of thousands came into use in Israel and Australia; the millions in Japan varied in type. There are few secrets except for some selective metal coatings which have so little additional merit in comparison with available black paints that they may be economic for special uses only.

As the price of oil, gas, and coal inevitably rises, a large market will develop for solar water heaters. Many heater designs are proposed and classified as water or air types with numerous subclasses of materials and configurations. It would be more useful to classify according to the levels of heat desired. As mentioned earlier, the human body is a solar collector consisting mostly of water in a semi-permeable membrane of skin — it prefers a temperature range of surrounding air to be 70-75°F to carry away internally generated heat. Thus, the body may be considered a limited collector or a selective absorber of ultaviolet radiation which normally rejects solar heat as well as an excess of ultraviolet rays (the latter it does by pigmentation). To a greater degree, the body is a heat dissipator. The most important frame of reference for solar use to produce thermal comfort for man is the thermomechanism of the body itself. A heating and cooling method which is based upon analogous principles will be mentioned last in this article.

Solar heaters used for swimming pools require only low-level heat supply since here, again, the desired temperature is below that of the body. The interest in swimming pool heaters results not only from rising fuel costs and the marginal value of a luxury, but because heated swimming pools are among the greatest fuel wasters modern technology has produced. They are highly efficient coolers in all except humid climates where they are sometimes such efficient heaters that methods for cooling, such as refrigeration, are used if affordable. Always one must come back to climate. The standard flatplate collector is used for pool heating, as is a black, extruded, plastic collector consisting of flattened channels; the latter, and in other cases coils of black plastic hose, are used without a transparent cover. A cover may create such high temperatures as to melt the plastic. The efficiency of solar collection by such simple collectors is greatly lowered by convection heat loss, despite less-radiation heat loss at the low collection temperature. A wind-break around the collector may be of some value, so long as plastic-damaging temperatures are not reached.

Swimming pools lose great amounts of heat by conduction to the ground and perimeter; the perimeter can be designed to minimize such losses by itself acting as a solar collector, though a high efficiency would preclude walking on the collector surface in mid-day. An effective but, occasionally uncomfortable solar collector may be a transparent enclosure over the pool or a more economic but somewhat inconvenient transparent plastic film on the water. These reduce convection and evaporation cooling while they permit daytime solar heating; by maintaining a high humidity in any air present under the cover, radiation losses are also reduced. Transparent plastic, itself, is not so effective in preventing radiation heat loss as is glass; hence, glass is used as the cover material for most flatplate collectors intended to produce higher temperatures.

Solar heaters for domestic water supply have lower efficiencies as the delivered water temperature rises. There is little need, however, for domestic water over 110°F except to reduce the size of the storage tank; a supplementary heater can provide higher temperatures when necessary. For household use, it is undesirable to heat water to more than 140°F, and most often, the hot water must be blended with cold water before one can use it. Unfortunately, most solar energy advocates attempt to duplicate the performance of fuel heat sources in an effort to meet industrial rather than human temperature levels.

The price for the most common flatplate collector is presently about $7.50 per square foot; this is expected to drop to the $2-4 range in the not-distant future. The roof-mounted collectors in most widespread use have been of the thermosiphon type which requires a water storage tank at a still higher level. This tank is unsightly if outside the house; it has often been camouflaged as a chimney. Whether inside or out, the tank provides an appreciable concentration of load near the ridge of a roof. Gravity feed of the hot water does not produce the flow pressure often desired. This restricts temperature increase by means of a conventional heater and prevents blending, in a faucet, with highly pressurized cold water. Solar collectors are currently proposed which operate under reduced pressure so as not to cause leaks in the collector; then they may have a pump to rebuild the pressure to that of the cold water; such a system requires control valves and adds operating and maintenance costs.

Many problems may be encountered with flatplate collectors. A long life has been obtained when the collector sheet and tubing are of copper and installed in an area where the water will not freeze at night. It was mentioned earlier that water can freeze by radiation cooling when air temperatures are much above 32°; this effect is reduced by a glass cover but radiation freezing should be taken into account. The scarcity and high price of copper is such that many collector manufacturers question the availability of copper for the large number of collectors expected to be installed in the near future. Stainless steel is also proposed; it would have the same scarcity problems.

Cost and availability favor the use of aluminum and steel for flatplate collectors. Despite more than 50 years of corrosion-control research by the metal producers, the federal government has been advised by personnel of large corporations who have studied solar collectors that much more corrosion research is required. This is true even though distilled or deionized water is used in the collector system and chemical inhibitors, such as those commonly put into automobile radiators, are added. There is no simple and sure answer to the corrosion of these metals, consequently, large manufacturers of aluminum or steel solar collectors do not give a long-term warranty.

To keep water from freezing and bursting the tubes of flatplate collectors, two techniques are used. First, the water may be drained if its temperature gets dangerously low. Drainage, however, introduces air which may accelerate corrosion of the collector and if drainage is incomplete, damage would result. As evidence of the seriousness of the problem, nitrogen gas is used to fill the drained tubes of one very large array of solar collectors. A second possibility is one of adding an anti-freeze, such as ethylene glycol, to the water circulating through the collector. This spawns other problems.

Ethylene glycol is fairly expensive. Certainly, it would be in short supply if required for a large number of solar heaters. It is toxic; this creates hazards of initial use and of later disposal. Careful protective measures must be taken to isolate the glycol heat-collecting fluid from potable water to which the solar heat is transferred through a heat exchanger. Two heat exchangers and multiple safety valves have been required for some installations. Ethylene glycol has a still unknown exposure life in a solar collector; its decomposition products are acidic and corrosive. If the antifreeze is not changed regularly, there may be collector corrosion and subsequent water leaks. It is easy to understand why flatplate collectors previously attained commercial success in areas where freezing was not a problem.

The installation of flatplate collectors also presents some roof leakage considerations, code requirements, and labor trade jurisdiction problems. These matters should not be more than temporary deterrents.

Some collectors are made with plastic covers to eliminate vandal or storm breakage of glass and the resulting hazard from glass shards. There are divergent views about the need for tempered or double-strength glass. The preference for single glazing, in southern states, and for double glazing in the middle and northern tiers of states, seems established. The optimum angle of inclination for the collector differs summer and winter; a compromise angle near that of winter optimum may be best. The angle becomes subordinate to esthetic considerations if a portion of the roof, properly oriented to the south or slightly west of south, has an angle close to optimum. The collectors may be mounted on porches, over windows, or as detatched, free-standing units in the yard.

Space heating with solar energy becomes more complicated when heat is obtained from flatplate collectors. The costly collectors are generally added over the roof structure and must cover a large area; 50% of the roof area is proposed for building in northern states and this is not expected to supply all the needs of a house. Greater coverage is not considered to be economic and would encounter constraints of shading, orientation, and esthetics. Focusing or concentrating collectors of the parabolic type have no advantage for space heating. The amount of energy available for collection is the same for a given collector area of any design. Concentrating collectors serve only to increase the water temperature — an asset if one wishes to have temperatures above the boiling point of water at the cost of lowered collection efficiency and higher initial and maintenance investments. In northern states, the conventional space heating system cannot be displaced by solar devices. Consequently, there is an additional cost for solar systems as well as associated charges for capital investment and maintenance throughout the life of the building. In southern states, flatplate collectors covering 50% of the roof may provide all or nearly all the needed heat and under favorable conditions not require a conventional heater provided that there is adequate heat storage to carry through days of low solar radiation.

It was formerly suggested that flatplate collectors should be installed with a heat-storage capability of three-days' carryover through cloudy weather. Recent analyses of costs indicate that frequently this capability is not economic. New recommendations suggest only two nights of carryover before supplementary heat would have to be used or a heat pump would be required to upgrade the low-level heat in the storage tank. The flatplate collector system (termed an active system because of its reliance upon circulation of the heated water) requires a heat exchanger to separate the ethylene glycol collector fluid from the water storage and distribution system which, in turn, calls for piping, pumps, controls and a network of radiant baseboard heaters throughout the house. The heated water must be considerably hotter than the desired air temperature for this type of radiant heating (referred to as a hydronic system). It is generally preferred to store water heated to 160°F or above for this type of system; thus, the piping and storage tank must be well insulated.

With the exception of the collectors, the items needed for the solar hydronic system are standardized, mass produced, and locally available items which will not drop appreciably in cost as a result of increased use. Hence, it should not be expected that the total system cost will drop 50% when the price of collectors are reduced by that amount; the total reduction from today's cost may be more nearly one of 20-25%. The active system has several critical disadvantages. The storage tank is buried in the ground, put in the basement, or it may take up floorspace elsewhere in the house. The tank is usually insulated against heat loss; heat lost to the room may provide needed warmth, at some periods, but unwanted heat at other times. The system always requires operable pumps; during a blackout, an active system would be inoperative just as would a conventional central heating system. (Repeated warning: consult your local utility about a guarantee against future blackouts.) A solar hydronic heating system, like conventional ones, has the disadvantage of high temperature radiation, hot and cold spots, wall smudges from localized strong convention currents, et cetera. So far hydronic heating can save energy; other than that it does nothing new except that in favorable regions, at higher first cost, it may provide a lower life-cycle cost for heating needs. Space heating and the domestic hot water supply are sometimes integrated — the heat being supplied by the same flatplate collector. Until forty-year old problems are solved, use of flatplate collectors may be restricted.

The Thomason Solaris System differs from most solar space heating systems in the form of the collectors and of the heat storage. The collector consists of roof-mounted sheets of blackened corrugated metal with a glass cover and insulated back. Recirculated water trickles down the corrugations from a perforated header near the roof ridge. The open streams of water absorb heat from the blackened metal and are collected and pumped to a storage tank. This tank is surrounded by small rocks for additional heat storage and so permits circulation of room air through the rockbed. In the Wathington, D.C. area, the Thomson system is reported to supply nearly all the needed space heat but a supplementary heat source is employed for the coldest and sunless periods.

The Thomason system is used for partial cooling in summer. The circulating water loses some heat at night, though radiation is obstructed by that same characteristic of glass which makes it a valuable cover for solar energy collection: low transmission of heat waves. The glass also reduces convection and evaporation; these cooling effects can be maximized by draining the water down uncovered corrugations on the north slope of the roof. Thomason prefers to cool the recirculating water with a small air conditioner operated at night for greater efficiency and to employ off-peak power. Several patents cover the Thomason systems which have been used on several houses for many years; unfortunately, adequate scientific data on cost and performance have not been reported. The scientific community has used this as an excuse to make little mention of this working system; the press has given the method abundant notice but as with all solar systems to date public acceptance has been slow.

Solar collectors using air as the heat transfer medium have long been studied for agricultural purposes, such as crop and lumber drying and for heating animal sheds. Air heat-transfer designs for the solar heating of houses may be as simple and passive as a greenhouse built against the south wall. Another type of construction uses a solar radiation absorbing material within a glass-covered box mounted outside a south window. Air entering the box near ground level rises as it is heated and discharges through the window opening into the room. Employing a similar, much larger collector, the heated air may pass through a bin of rocks from which it is transferred into various rooms by natural convection. A French design admits solar radiation through a vertical facing of glass and through an air space before it is absorbed by a thick, blackened concrete wall which stores the heat and releases a portion of it to air rising between the vertical surfaces of the wall before discharging into the house.

In a more complicated (active) system, air is forced between multiple layers of glazing in a roof-mounted solar collector. The heated air then passes through a bin of rocks for storage of heat. Recirculation of room air through the rock bin provides warmth when the collector is not operating. For 15 years, this installation has supplied about 26% of the energy for space heating and a portion of the domestic hot water in the house for which it was designed. In another installation, the choice of an air system appears to be based on an extraneous reason — cadmium sulfide cells within the collector must be cooled by the air stream. The air forced from the collector of this installation transfers heat to salts stored in plastic containers. This storage method depends upon the melting and subsequent recrystalization of the salts; but 30 years of research have not adequately overcome difficulties with recrystallization. To date, it has not been possible to get the salts to change phases enough times to be practical for use in the space heating of buildings. A system which uses three different salts for differing effects seems to compound the potentials for trouble.

Air is a poor heat transfer medium — a low-efficiency one. This, alone, should not be regarded as an obstacle unless other disadvantages of an air system make it less desirable than active systems based upon water circulation. The volume of air flow through the collector, ducts and heat storage medium must be large; due to friction in these parts, a large fan and motor are required for circulation of the air. The noise level and discomfort of uneven air temperatures in various parts of the house are comparable with those of forced-air furnace systems. Some passive air systems are restricted to rooms of southern exposure. Passive systems, however, have considerable potential for development through imaginative architectural design.

Solar collectors are often criticized as being unesthetic; this is a generalization which does not apply to all important categories of collectors — but generalizing has always been more convenient than the learning of processes and the acknowledgement of broader truths. Flatplate collectors, as used in northern states, are conspicuous because of their steep angles and the new appearance of large glazed areas on the southerly side of a roof. One suspects that they are particularly unesthetic for manufacturers and installers of other roofing materials or heating systems. Greater dependence upon solar energy will convert this appearance into a sought-for asset. The steep angle with uniform southerly orientation, however, may always present architectural constraints and may result in some wastage of living space. If loss of living space is caused by collectors or heat storage, this loss should be a charge against the solar system; not all systems produce a loss of space.

The future of solar energy for space heating and cooling will, primarily, develop through improved architectural systems rather than through mechanical systems similar to conventional appliances for thermal comfort. Most solar scientists are ill-prepared to think in terms of, and are unwilling to work in collaboration with, architects as team leaders. A few examples will indicate where solar scientists have gone wrong by concentrating on mechanical systems.

### Where Solar Research Went Wrong

The effectiveness of the national program for solar energy development, up to 1975, is open to serious criticism. This is not to impugn either science or research. Recognition is warmly given to all those meritorious efforts that have led to the collection of basic information on selected aspects of solar energy. Valuable research is, unquestionably, being done that will, in the future, benefit solar energy's use worldwide. Yet the solar program, through 1974, appears to have caused greater harm to the early commercialization of solar energy than it has been helpful to it.

The basic error of federal solar energy research for the heating and cooling of building was its initial concept. Failure to begin a state-of-the-art study with a comprehensive analysis of working systems and field data resulted in overemphasis on mechanical (active) systems and disregard of passive and architectural systems. The consultants who were used were best known for many years of engineering research that had minimal acceptance in the marketplace and in filling national needs. Long-term research proposals, largely restricted to the interests and capabilities of the consultants or advisers, were presented as the recommended national policy for solar energy development. Engineering theory and computer analyses were regarded as the high technology assurance of success. Institutions that depend upon endless research by people not noted for practical accomplishment were favored with big contracts. Impressive documents and "sideshows," which in many cases seemed motivated more for political and appropriation support than for practical results, were objectives. A few specific cases can be mentioned.

High cost, uneconomic demonstrations were made of solar heating four schools in northern states. For more than 30 years, the scientific literature had made clear that solar energy would have difficulty attaining success in this region where most of the research has been done with the type of flatplate collectors chosen for the school demonstrations. There was little new technology involved, but there was a wasteful rush to spend money at a time when Congress was critical of past inaction. The obvious conclusion was disregarded. The southern tier of states had long been characterized as having "maximum feasibility"; and the northern tier as having "minimum feasibility." True, the **need** to conserve fossil fuels by use of solar energy is great in the region of the demonstrations. Winter temperatures are low there precisely because there is so little sunshine; this provides a poor base for economic use of solar energy. Under subsequent pressure, the federal demonstrations included a school in Atlanta, Georgia.

Officials in charge of solar research at the federal level said they would **not** study solar energy in the southwest because its use there was "too easy." Yet, there was no commercialized system for solar heating of buildings in this area of greatest solar radiation in the United States where vast quantities of gas and oil were being consumed. Nor did the officials identify their system which would make solar heating "so easy." They were familiar with one passive system which had proved its potentials for 100% heating and 100% cooling in the region, but no mention of this method was made in the agency's reports nor in its statements to Congress.

Consistently, the federal agency stated that three years of research would be necessary to develop solar heating of building and five years for **solar** cooling. This schedule was accepted by Congress and the press — consequently, by the public. The research was expected to produce unspecified new developments; at best, this was a hope without assurance of practical realizations. A mentality designed NIH (not-invented-here) is not unknown in large companies and often applies to bureaucracies and to their consultants. An idea has little merit if not developed within the organization involved. In this case, the successful but unmentioned possive method was being evaluated by a government agency other than the one having main jurisdiction for solar energy. Obtuse justification for not referring to the cooling accomplishment of the unmentioned method was that it was accomplished by **nocturnal** cooling and not by **solar** cooling. The major goal of the President when he gave the agency jurisdiction for finding solutions to the energy crisis was to conserve fossil fuels and electricity — a goal which should be above bureaucratic semantics. Wind power, biothermal conversion, ocean thermal schemes, et cetera were acceptable to the agency as long-term projects; but nocturnal cooling, which had immediate application potentials and was combined with solar heating, was not furthered or discussed.

It is justifiable, then, to look at the **solar** cooling method which got strong support and was called to the attention on Congress and the public. Under a $238,000 contract involving two universities, a solar cooling system was designed to use a three-ton, lithium bromide, absorption cooling unit, described as being of the type that is "the only commercially available cooling device suitable for solar operation. . ." This unit was expected from computer studies to provide approximately "88% of the cooling load in a randomly selected prior year." The absorption cooler was used in conjunction with rather standard flatplate collectors which were to be more economical through provision of approximately 82% of the heating load. First year performance indicates about 65% cooling and somewhat lower expectancy for winter heating, hence, the system has far less prospects for being economic than were expected. The cooling result was most disappointing considering the favorable cooling conditions of the site.

What is more disappointing in this solar cooling demonstration is a major flaw that would have been detected by a high school science student. To lower the condenser temperature of the 3-ton absorption cooler which cost $2,000, an 8-ton evaporative cooling unit costing $550 was added to the system. Together, at a cost of $2,550, these components produced only 65% of the needed cooling. Put another way, they performed only one-third so well as might be expected from the $550 component alone — without complicated controls and without solar energy. If this rather specious experiment were clearly identified as an evaporative cooling system handicapped by solar energy complications, there would be less touting of it as the first demonstration of solar cooling. It is hoped that other demonstrations of solar absorption cooling will not be subject to this criticism.

Another indication of major expenditures producing little progress for the solar heating and cooling of building is evidenced from contracts of over $500,000 each to two large manufacturers of conventional heating and cooling equipment and a like amount to an aerospace company. These companies were to come up with new concepts to be further developed with larger contracts. The reports submitted in fulfillment of the first contract contain much useful data, but no basically new solar heating and cooling methods. More discouragingly, the costly mechanical methods proposed were not expected to make an appreciable impact on the nation's energy consumption for heating and cooling until the year 2000. The recommended route for this slow development of solar energy is government subsidy of high-cost systems on government buildings until about 1985; then, it is expected that some commercial and industrial buildings might be equipped with the systems. There was little hope that residences would be heated and cooled in anything like the goal sought by Congress.

Although one of the three reports reached favorable conclusions about nocturnal cooling for Los Angeles, Phoenix, and Washington, D.C., it did not recommended further demonstration. Another report acknowledged that solar heating and cooling would be easiest in Santa Maria, California, but the proposed demonstration program for the area would build only 16 houses with solar heating, 24 with solar assisted heat pumps, and 16 with solar heating and cooling. In contrast, in a northern city of "minimum feasibility" for solar energy, the demonstrators would build 90 houses — 60% more than in the area of greatest feasibility. This becomes more inexplicable in light of the report's figures showing that solar heat can complete with conventional fuels in Santa Maria in 1975, but not until 1985 in the northern city (Madison, Wisconsin). The solar assisted heat pump and solar heating and cooling were expected to be competitive in Santa Maria in 1980, but in Madison not until 1985. This was the fastest performance expected by this company.

But less than 50 miles away from Santa Maria, in a more severe climate, a house was operating with 100% solar heating and 100% nocturnal cooling with better economy than that which can be expected with the mechanical systems. This successful passive system, widely described in the technical literature, was known by the company which recommended to the government the study only of complicated mechanical systems which might someday fit into their production lines. It is possible that solar energy development would have proceeded along such lines, if Congress had not reshaped the federal approach.

Pointedly displeased with past accomplishments in commercializing solar energy, Congress ordered early demonstration of solar heating and cooling of buildings and placed the National Aeronautics and Space Administration (NASA) and the Department of Housing and Urban Development (HUD) in joint charge of a large program. HUD is to further residential use of solar heating and cooling while NASA will apply it to other types of buildings. By definition in the law, Congress specified that the term solar cooling is understood to include nocturnal radiation and other non-solar methods for reducing peak load power demand.

Politicians are acutely aware that no easy answers exist for a continuing and worsening energy crisis. At both federal and state levels, legislators are ahead of other segments of our society in practical promotion of solar energy. On the whole, solar research experts are not partial to legislated acceleration; they say that there is much research yet to be done — which is true for a Cadillac or the best TV set; they say that an early demonstration will freeze technology — which is a plea for research forever; and they say that there are too few competent professionals to guide accelerated programs — which means that they want to maintain their own monopoly position. If deductions were made from research budgets of costs to the public resulting from delays unnecessarily caused by research, researchers would be greatly in favor of accelerated programs.

The consequence of having research people slow down solar energy use is that each day about 5,000 houses are completed (and are being planned for future completion) with reliance for the next 40 years, or more, on conventional heating and cooling devices which depend on gas, oil, or electricity. There is no assurance that 10 years from now these forms of energy will be available in the amounts needed or that they will sell at affordable prices for thermal comfort. The delay resulting from carrying out millions of dollars of long-term research, instead of accepting proved solar systems can cost billions of dollars worth of oil. In addition, millions of houses completed during the research delay will have conventional thermal control systems which will be obsolete and which will depreciate the value of the houses and of the mortgages on them. This financial loss to the public is a price for long-term research and slow progress in the development of solar energy which exceeds toleration. Congress is highly justified in demanding faster action from solar scientists and from government agencies.

### Changes Ahead

Because of natural gas and oil shortages as well as higher prices for coal and electricity, the public looks anxiously for clues concerning the inescapable changes ahead. Minimum change is wanted; solutions should not lower living standards; anything truly inconvenient should be for future generations. Unfortunately, vested interests are concerned with even minimum changes now; in their charges and countercharges, they create public confusion. The issues become larger than can be assimilated without arousing collective emotions. The ideal path would be one of going back to familiar technology provided that a serious lowering of living standards did not result. Happily, this is possible for the heating and cooling of building in at least some locations.

A change already made by many young people is an abandonment of faith that high technology will come up with miraculous answers. What is sobering is our realization that such answers created our present problems. What is even more sobering is that our experts can be so wrong continuously. Here is an explicit example: federal energy officials now propose replacing gas and oil heating of buildings with electric resistance heating; the California Coastal Zone Conservation Commission proposes to ban resistance heating because "Use of electric resistance space heating results in consumption of at least twice as much energy to heat a given space as direct use of a primary fuel (e.g. gas or oil)." The federal experts, presumably, continue to place faith in the future of coal and nuclear power plants. The environmentalists see these plants in terms of more strip mining and pollution. With no major solutions apparent, both extremes attempt less to find truth than to find an acceptable compromise which will ameliorate the present crisis. Even compromises will require great changes.

The American Society of Heating, Refrigeration, and Air Conditioning Engineers has prepared a 40-point checklist for reducing energy demand within a house. Some of these energy saving changes cost nothing and create no inconvenience, for example, not heating and cooling unused rooms. Other changes cost nothing, but they create frequent inconveniences throughout the day, such as raising and lowering thermostat settings, turning off unneeded lights, opening the refrigerator and freezer less often and as briefly as possible. Nominal expenditures are required to follow the ASHRAE advice to have the heating system cleaned and adjusted each year and to have furnace filters replaced every 60 days. A few recommendations necessitate larger expenditures for installing storm windows and for adding insulation. ASHRAE suggests that the public pay more attention to solar energy: to open shades and curtains to let in sunshine during winter days and to cover windows at night (in summer, the reverse procedure should be practiced, but people dislike losing privacy and exposing interiors to the view of prowlers); to shade air conditioners; to install awnings and sun screens; to plant appropriate shrubbery or trees, strategically. All the ASHRAE proposals involve more study and thought on each use or planned use of energy than has recently been required of a householder. Just to learn and practice energy conservation measures is going to be a major change for the general public and one that brings attention closer to the effects and potentials of solar energy.

Planners and architects are proposing many changes in and around the house; these professionals can no longer supinely let their client's present wishes dictate continuing poor practices for which the professionals will be blamed in future years. The major changes call for smaller streets and houses; grouped housing with common walls; privacy fences that allow south-facing windows to have curtains open; smaller windows where solar heating is not involved; exterior movable shutters of insulation; and use of ventilation cooling. To absorb the solar energy passing through windows, concrete floor slabs are advised — carpeting is to be minimized. Patterned concrete or exposed aggregate, ceramic tile, or vinyl coverings are recommended so that the heat capacity of the floor can be credited in some proposed codes which specify a proper ration of internal heat storage to south-window area. Masonry construction will be stressed despite higher initial cost and particular emphasis will be given to water storage within the house to modulate temperatures.

One can, of course, anticipate loud choruses of complaints from those who will spend more time drawing draperies, walking on uncarpeted floors, and changing thermostat settings as well as remaining at home to let in workmen who will clean and adjust the many energy-consuming devices. Some householders will more strongly object to the thought that improved paths for bicycles and walking are meant for them or that the provision for old-fashioned outdoor clotheslines means they are to be used. Some life-style changes will produce initial reactions that will be modified, of course, by the increasing cost of supplied energy.

Polls taken on expressed attitudes toward giving up certain appliances will indicate that those appliances which people are most ready to stop using are small consumers of energy, while those given up reluctantly are responsible for the large energy drains. Air conditioners, freezers, dryers, and automobiles are still rated essentials, but there will be little energy saved unless their use is restricted. Ignorance of the ease with which air conditioners can be avoided and unawareness of the greater comfort possible in doing so cause the rapid increase of air conditioning to be the most unconscionable strain on electric power supplies. This waste of electricity in the house deprives industry of its power supply and also causes unemployment and shortage of goods.

The major reduction of energy use in buildings must be for their heating and cooling. In some regions, space thermal control accounts for two-thirds of residential energy demand; another 15% may be attributed to water heating. These are precisely the areas for which solar energy and nocturnal cooling are developed to the point of practicality. In some climates, the natural energy forces are so great as to provice an excess for both heating and cooling. Under such circumstances, the diurnal energy forces can be wasted with impunity, if the system using them does not increase initial costs of housing — a distinct possibility.

Evaporative cooling has long provided thermal comfort in arid-region housing. The evaporation of one and a half gallons of water per hour is equivalent to the cooling effect of a one-ton refrigeration air conditioner, the operating cost for the evaporation unit is far less. Those recommending large scale use of evaporative cooling should, however, acknowledge our increasing water shortage. There is little advantage in solving one problem by creating another one — as too often has been our blithe practice. River water polluted by sewage or industrial waste entering a few miles upstream is the only drinking water source for many cities. Rivers are channeled hundreds of miles through aquaducts without irrigating vast areas of desert land through which the water flows to cities and houses. At otherwise favorable sites, water is becoming unavailable for thermoelectric generation needs. Production of synthetic gas from coal is going to be limited by water availability near coal fields. A kindred water problem resulting from strip mining of coal for electricity is the deleterious effect of rain leaching salts from the newly moved overburden. Streams will be made increasingly saline and useless for irrigation; the United States already is paying Mexico damages for the effect of the high salinity of the Colorado River leaving our country.

Evaporative coolers frequently produce a brine which corrodes the equipment and contributes to a pollution load upon discharge. Evaporation is not effective during humid periods which may be combined with 100° temperatures in some parts of the United States. The most used evaporative coolers are active systems with electricity-consuming pumps and blowers that are also major repair items, along with filter replacement. For these reasons, passive evaporation has merit, if evaporation cannot be avoided by use of nocturnal convective or radiative cooling. The federal government should publish a map showing those regions of the country where the various cooling systems can be used advantageously to eliminate electric air conditioners. The map should also indicate where solar heating is economic with today's technology. Such a map would show that wherever solar heating is not economic, because of low solar radiation, there is little real need for air conditioning of residences and of many other types of buildings. If northern states are not blessed with sunshine, they certainly can be thankful that nature can provide free cooling. This map should reach the public soon, if the government expects cooperation in energy conservation. Delays in publishing it will result in continuation of large investments made needlessly for appliances which, when most needed, will have to be shut down due to lack of operating energy.

People are asking, increasingly, about the potentials for adding solar heating or cooling devices to their present houses. Perhaps only 25% of existing houses can be retrofitted economically for solar heating, except through use of the partial effect of the simplest passive systems. For flatplate collectors, roofs may not be properly oriented to the sun or they may be shaded by buildings, trees, or hills; the roof slope may not be optimum and may require collectors to project and to assume unsightly angles; or the house may be too old to justify such a high-cost system, unless a major renovation, such as reroofing, is contemplated. On the other hand, retrofitting for cooling by ventilation or by evaporation may be quite feasible using off-the-shelf items which can be installed by the same people now installing refrigeration air conditioners.

Maximum economy results when one system can do both the heating and cooling and when such a system is incorporated into the design of a new building. The greatest need is to have, in different parts of the country, side-by-side demonstrations of all systems which claim to be practical and economic, so that local architects, builders, and homeowners can compare the systems and choose one on the basis of cost, performance, and appearance. This comparison could be sponsored by city or county governments, with only small subsidy to buildiers. The city of Colorado Springs went further by forming a non-profit corporation to demonstrate one solar system; a different system is used by the private builder of a house across the street from this demonstration; thus, with a few more systems built in the area, the basis for comparative testing exists within that city.

Various states have started to encourage the use of solar energy. The California Coastal Zone Conservation Commission's actions toward forcing developers to consider solar heating and nocturnal cooling before issuing a building permit are the most firm and positive ones; a Florida law requires that all new buildings be designed for future installation of solar water heaters; and Arizona law allows a state tax deduction on solar equipment amortized over five years; Indiana's law exempts solar systems from property tax assessment; Ohio's new energy budget code for buildings will stimulate interest in natural heating and cooling; and Wisconsin's new electric power rates remove the incentive to use larger quantities of electricity. Other states and Congress are considering laws which provide a great variety of incentives for the installation of solar devices. These incentives include several forms of tax relief, low-interest loans, guarantees on mortgages and insurance, assistance to manufacturers of the solar devices and to architects who expend extra effort designing for optimum use of solar energy. These incentives will, assuredly, be expanded until the objective of reducing fossil fuel and electricity consumption for thermal control in buildings is attained.

### A Promising System for Solar Heating and Nocturnal Cooling

Far fewer articles would be written about solar energy if, as a prerequisite, the author had to justify the reader's time by disclosing something of proved, practical, and economic value for a prescribed portion of the United States. Ideas which are for the distant future, based on theory, not proved, not acceptable, uneconomic on first or life-cost basis, or that are not specifically indicated as having geographic or climatic limitations all tend to cause the public to lose interest in solar energy and to turn away from other systems that warrant consideration. There are several meritorious systems in operation today. One is unique in having proved its capability for 100% heating and cooling in the Southwest. This system is likewise unique in being impartially tested and evaluated in all details of construction, thermal performance, and occupants' reactions. Prototype flaws have been publicly disclosed and modified so that a production design will provide maximum satisfaction with its cost, performance, maintenance, and acceptance.

This impartial evaluation was financed by a small project funded by the U.S. Department of Housing and Urban Development and is the only one to scientifically study the performance of a solar building throughout year-round normal occupancy. Eight professors from California State Polytechnic University, in San Luis Obispo, California, have studied the prototype for the past eighteen months. The house was built to the professors' specifications, but construction costs were entirely met by the writer who has patents on the system, known as Skytherm Southwest. A quite different design, using some of the same principles, but adapting the system to snow country and to the lower solar intensity of northern states, is designated as Skytherm North. With patent protection in progress, Skytherm North will be built and then publicized after adequate proof of the performance characteristics with, hopefully, an evaluation as complete as that now described for Skytherm Southwest.

By applying the principles of natural heating and cooling mentioned early in this article, it was found that ancient building technology could be combined with a minimal amount of modern technology to produce a passive thermal control system having comfort potentials exceeding those of conventional appliances. From what has been said earlier, it is clear that solar heating should be coordinated with nocturnal cooling in a building, just as it is in nature. Not all the old Spanish Missions, in California, are comfortable inside on summer night; nor are they all comfortable without supplementary heating on winter days or nights. The new technology that produces 24-hour comfort from a varying diurnal energy flux is that of movable insulation operating in consonance with the energy levels and in combination with heat storage.

Skytherm Southwest was first tested, in 1967, in Phoenix, Arizona. The results were superior to any thermal control previously obtained with natural heating and cooling. The test room was kept within American standards of comfort during both dry and humid periods when outside temperatures ranged from freezing to 115°. The present evaluation of the system by the professors was to determine whether the same conclusions applied to a full-scale, occupied house in an area having colder and less sunny winters but cooler summers (temperature range 17 - 100°). The system has now been tested in a severe desert climate and in the moderate, valley climate of Atascadero, halfway between Los Angeles and San Francisco.

To use heat storage principles, the house was built with a concrete floor slab and with concrete block walls running east-west. Over the walls was placed metal roof decking which forms the ceiling of the rooms. Above the metal is a plastic liner comparable with that of a swimming pool. Over this, and between beams, are plastic-enclosed bags of water 8 inches deep which provide more heat storage for the entire living area. In trackways mounted on the beams, are panels of plastic-foam insulation which is spread across the water bags or is pulled to a three-deep stacking position over the carport, utility room, and patio by a small motor running only six minutes per day, during the opening and shutting operations. Because concrete construction is more expensive than is wood frame design in California, the north-south walls, containing windows and most exterior doors, were of frame construction with standard insulation.

In winter, the insulation panels are stacked during the day to allow solar radiation to heat the water to 85° — a much lower temperature than that sought in flatplate collectors. Heat from this water is transferred through the ceiling to the inside walls and to the floor; these then gently warm the room. At night, the waterponds are covered by the insulation to prevent heat loss to the sky while the water, walls, and floor give up a part of their stored heat. All rooms can be kept comfortable through three sunless days. If, after the three days, the house becomes slightly cool in the morning (as occurs with conventional heating when the thermostat has been turned down), the room temperature will recover gradually while the panels are open to recharge solar heat in the ponds. The few degrees greater heat in the room is usually during the evening hours when the family is less active and enjoys a bit more warmth. The occupants of the house expressed the opinion that the heat felt more natural. This reaction, in part, may have been a consequence of the minor temperature swing and the normal humidity that contrasts with the dry air resulting from conventional heating. The occupants reported winter comfort as "distinctly superior" to that of gas heating.

In summer, the opposite movement of the panels exposes the water bags at night to discharge, by means of radiations and convection, the accumulated heat which had infiltrated or had been generated into the rooms. At Atascadero, it is not necessary to flood the water bags with a layer of water to obtain the evaporative cooling that was required in Phoenix. The rooms were all kept in the 69-76° average temperature range, when outdoor temperatures during summer months reached 95°. The family reported summer cooling to be "far superior" to refrigeration air condition. The reason for this was again that the humidity was normal and that there were no hot or cold spots in the room; also, there were no noises or air drafts.

Because of unusual use of water on the roof, people have certain questions about the system. These can be answered as follows:

1. Water ponds must cover 100% of the roof area in a severe climate; in mild climates 100% coverage may be enough for a two-story building. Because of the low cost of the roofpond design, it may not be economic to have only 50% coverage of the house; it may be cheaper to reject an excess of thermal potential in the Southwest.
2. The water depth may be varied according to the climate; there is more comfort (as well as more days of carry-over through cloudy days) the deeper the water is. The 8 inches of water used, in Atascadero, weigh no more than a four-inch concrete roof deck. The design passed the earthquake code for an area having a major geologic fault 30 miles away; during the evaluation, the roof was unaffected by a nearby quake of Richter Scale 4.8 intensity.
3. The water is retained by several impervious membranes (the metal deck, the plastic liner, and the water bags formed like waterbeds). The water in the bags cannot freeze because, when not being heated by the sun, the bags are covered by insulation. If the house is left unoccupied for weeks during winter, freezing would be delayed by heat given up from the walls and floors. Even if the water froze, the plastic bags would not be damaged.
4. Algal growth in the water can be controlled by chemicals; mosquitos cannot breed in the closed plastic bags (no water is exposed when radiative and convective cooling are adequate). In areas where evaporative cooling is necessary, a chemical treatment may be added to the water flooded over the plastic bags, but shortness of the period

Integrated concrete block and wood construction shows versatility of materials and architectural effects. Results are best with new concrete or metal construction, which may be conventional in appearance. SKY THERM is not limited to the one-story, flat-roof style shown; designs are under development for southwall heating in northern and snow regions.

SKY THERM provides new, more comfortable standards by mildly heating and cooling through an acoustical metal ceiling - no hot or cold spots - no air drafts or noise. Regular building materials <u>are</u> the heating and cooling system; hence low initial cost compared with conventional design.

when evaporation is needed may make chemical treatment unnecessary.

5. The plastic bags will have long-life because they are shaded by the aluminum-faced insulation panels throughout the summer days and are exposed to the deteriorating effect of ultraviolet only during a few winter months, when ultraviolet intensity is low. Urethane insulation placed inside buildings presents fire hazards not existing with externally mounted insulation. If the roof panels caught fire despite the aluminum cladding, the waterpond would effectively prevent fire and heat transfer into the house. An insurance official indicated that these factors would be justification for a lower insurance rating than for most of the construction used in California.

6. No supplementary heating or cooling is needed in the most populous areas of the Southwest. In colder areas, additional heating can be provided by extending the roofponds onto the overhangs or by use of windows and the vertical south wall as solar collectors. Emergency heating can be accomplished by leaving the lights on or by using the stove or other resistance heating appliances.

7. The system is automated, so that no decision need be made about proper timing to move the panels. People, generally, underestimate the heating potentials of an overcast sky. Light colored, thin clouds may have no effect. Broken white clouds may, temporarily, either increase solar radiation or slightly reduce it. Only heavy, dark-grey, complete overcast which passes too little solar radiation to cause a shadow is so low in energy that roofpond water cannot be heated above 70°, when the ponds are in contact with air at 32°F. A shadow is a good measure of enough solar energy to replace gas or oil heat.

8. Horizontal roofponds are not unesthetic — they can't be seen unless the architects have used a stepped-up pattern for an appearance other than that of the indigenous, Southwest style.

9. Flat roofs are used on most low and high-rise buildings throughout the country (not on most residences, but on a substantial percentage of them); these multistory buildings can have their upper story heated and cooled by roofponds and by moveable insulation. Any excess thermal effect can be transferred to the floors below by a water circulating system comparable with those of other solar hydronic systems. In California, it is now expected that most two-story buildings with light heating and cooling loads could receive most, if not all, thermal control from roofponds; 3-story buildings might be fully heated and cooled most months, but the first story might need supplementary heating and cooling during mid-winter and mid-summer months.

10. All solar systems are limited in use to unshaded areas; houses must not be built under trees nor have large trees to the south side. At two installations, blowing leaves presented no problem for roofponds. The leaves usually blow near ground level. Dust has little effect on solar collectors — less on nocturnal cooling; dust can be washed off as a part of a routine servicing contract if necessary. Solar cooling systems have the same shading restrictions as those which apply to solar heating; nocturnal cooling does not because the cooling occurs to the zenith sky at night.

The most frequent and most important question about a solar system is its total installed cost. Solar heating and cooling systems are far from being cataloged or off-the-shelf items such as window air conditioners. Flatplate collectors of standard types have relatively fixed, regional f.o.b. prices and may eventually be stocked by retailers, such as hardware stores and heating-ventilating air conditioning companies. The total system, however, requires many other parts not so readily stocked, such as the large storage tanks for solar hydronic system. Roofers may be needed to work with plumbers on a hot water installation; excavating machinery is required, if the storage tank for a heating system is to be buried in the ground. Solar systems are too complicated for installation by most individuals, hence, their costs will be variable in different regions.

Architectural systems have additional variables when the total building becomes a part of the system; the walls and concrete floor may well be considered a part of the system, though, eventually, it may be found that all buildings will again be built with high heat storage materials to conserve energy and to provide greater comfort. The cost differential of frame versus masonry construction is a factor in an architectural system and compromises may be made as in the combination of the two for the Atascadero house. Building materials and labor costs which may vary 10 to 20% within rather short distances can become critical in the design of an architectural system.

The ceiling and the roof of the roofpond system are also the heat collector-dissipator-storage-transfer means in many cases. Different structural and architectural effects vary the cost; the ceiling appearance may be varied (flat or ribbed); acoustical treatments may be added and become, in effect, a part of the heating and cooling system. Resistance to sound transmission can be a very important asset of the architectural system in areas near an airport, freeway, or factory. When it is not feasible to put the roofpond-type unit on a roof, it can be installed on the ground and the heated or cooled water pumped to a radiant heating or cooling installation in the house.

In addition to these considerations, architectural systems do not represent the rather static technology of the flatplate collectors and they will have considerable room for reduced cost as the parts get into mass production. The near-term goal is not only lower cost than a flatplate collector but zero first cost. This possibility is indicated in the reports of the California Polytechnic University professors as realistic and competitive now with the cost of a ceiling, roof, and heating and cooling system for a custom-built house. The General Electric Company has reported to the federal government: "The basic materials and design aspects (of the Skytherm system) should support the low cost potential claimed." Actual quotations for the system would have to await establishment of franchises at the local level and would be subject to installation variations.

The foregoing description of Skytherm Southwest presents an example of successful natural heating and cooling provided by an architectural system working in harmony with the climate rather than one of trying to overpower it. It has proved that a selection can be made of that part of the day's climate which gives one comfort and that part which does not can be rejected by means as easy as pulling a rope twice a day to move panels of insulation. Though this sounds simple, the principles and the technology which make the system work, as well as the design and installation of the roofponds are quite involved. In time, the system will be further simplified for installation by builders who will not have to be specially trained or supervised.

This discussion is an effort to show that simplification must come through an interdisciplinary approach. The disciples of high technology, on the other hand, seek to simplify by making a complicated mechanical device omnifunctional for thermal comfort (the heat pump being the prime example) and they nearly succeeded, until the burden placed on nature became self-expressive. There will be no breakthrough by engineers or architects alone or in combination. Meteorologists, biologists, chemists, physicists, and many other specialized professionals must contribute to a viable system based on incoming and outgoing radiation of natural energy. Through the interdisciplinary approach, we can solve problems which baffle us today.

### What Now for the Public?

The public quite clearly wants a simple solution to what it regards as a simple situation: more power resulting from flicking a switch. There are no simple situations to be improved; if they were simple, they would have been improved upon thousands of years ago.

Over the metal ceiling, a double, impervious plastic liner is under these waterbeds; above are panels of movable insulation. In winter, the uncovered waterbeds are solar heated; automatic closing of the panels prevents nighttime heat loss. In summer, heat absorbed from the room is stored in the water until the panels, which then prevent daytime solar heating, move aside to allow nightsky cooling. Phoenix, Arizona, tests established that SKY THERM Natural Air Conditioning required no conventional heating or cooling when air temperatures ranged from subfreezing to $115^{\circ}$. This result is unequaled by any other system for solar energy use.

A $\frac{1}{4}$ hp motor running only two minutes morning and night (or manual operation) moves the insulation panels (thermal valves) on trackways from over waterbeds to a carport or patio area for 3-deep stacking. In some climates, a little more electricity may be needed.

We must seek simple solutions to complex situations. We have seen that this does not necessarily result from large expenditures to develop complicated devices. The public can become a slave to the complicated device designed to serve it, whether the device is an automobile or means for thermal comfort.

It will only be:

After an intensive reeducation program to encourage acceptance of a new attitude toward energy use — that of moderate rather than maximum consumption;

After industrial research has changed its course from one self-centered over-confidence to one of vigorously seeking outside ideas (even critical ones) and pursuing those which best conform to the working principles of nature;

After government officials seek consultants with an established record of creative changes which survive the test of time and after those thinkers whose accurate analyses were rejected years earlier are rehired by the less far-sighted officials;

After man learns that he must re-orient his industrial economy and his physical well-being toward solar energy; that there can then be hope for energy use which will provide progress without retrogression.

After the millenium? No, soon. The crisis is so great that vendors of fossil fuels propose stripping the world of its remaining reserves — expecting government approval on the basis that an industrial collapse must be prevented while research comes up with new technology which will give the public any number of "energy slaves" necessary to satisfy the quest for power. The approach that has failed will be given more impetus and it will fail faster as a result.

The seeds of change are germinating. A minority of the public is already asking for stringent means to conserve energy. Industry is rapidly being pressed to the realization that it has the principal burden for introducing solar energy. A new government agency, ERDA (Energy Research and Development Agency), is being formed to coordinate the development of "alternative sources of energy." Government responsibility and officials are being reshuffled under pressure from politicians responsive to the concern of the public and to their own concern in finding that neither industry nor federal agencies have near-term answers to the sudden "shortage of energy."

Shortage of energy? The sun has sent and for millions of millenia will send to earth so much energy as to make that generated by man's advanced technology puny in comparison. The sun is our main source of energy — all other sources will always be the alternative sources. The public at large must gain a clearly perceived conviction of this truism. Then, it will be for the public to insist that at all levels of civil government, from the village to the Presidenty, every opportunity and power of persuasion shall be used to change practices which consume vast quantities of fossil fuels under the same roofs that are not built to utilize the solar energy falling on them. Individuals should not be expected to learn by extremely expensive trial and error methods which of the solar systems is most suitable for their use; in every county, there should be demonstrations by government of the most practical and economic diurnal energy heating and cooling methods. This much help government should give to insure the future welfare of man on earth — and a worth-while present.

*Late Note:*
*NASA no longer has a separate part of the solar energy program. The Energy Research and Development Administration (ERDA) will have a division dealing with the "alternate sources of energy" and will operate through the Department of Housing and Urban development for housing demonstrations and many activities affecting the acceptance of solar systems.*

# The Solar Tempered House

It is possible to get some of the benefits of solar heating without investing in solar collectors, pipes, pumps, storage tanks, etc. Most houses are built with total disregard for the sun. To make up for the drawbacks of the house, over-sized energy wasting heating and cooling systems are installed. A house whose elements are properly arranged is much easier to heat and cool. A solar tempered house is a house designed so that the sun works for it rather than against it. It looks quite ordinary. Its special properties are reflected only in lower fuel bills.

The main feature of the solar tempered house is a row of large south-facing windows under wide overhanging eaves. The eaves shade the windows from the high summer sun, but let in the low winter sun. The windows are double paned for minimum heat loss. During the winter nights heavy white curtains are drawn across them to help keep heat in.

During the summer trees help to shade the house. In the winter after the leaves have fallen, the sun shines through the bare branches to help warm the house. In the same way, vines on trellises help to shade the east and west walls of the house during the hours of the summer when the sun is low in the sky.

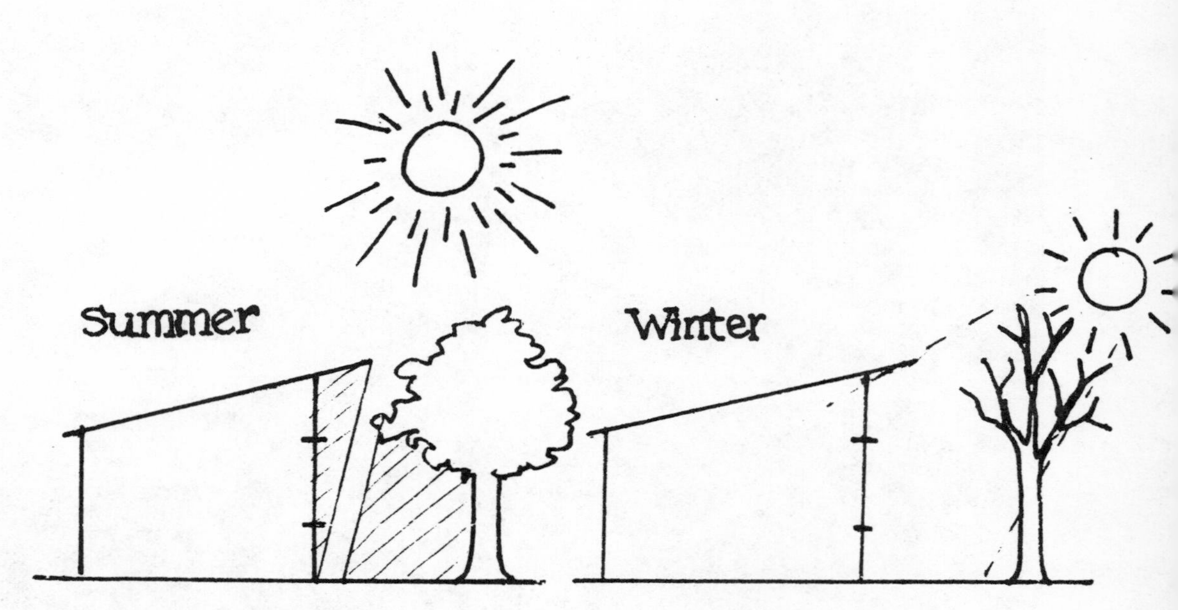


The roof is white because white objects are good heat reflectors and poor radiators. In summer, the roof reflects a large proportion of the heat striking it. In winter, when the house is warmer than its surroundings, the white roof loses less heat to the outside.

A solar tempered house costs no more to build than an ordinary house. It requires only some extra attention to design and layout. If you are planning to build or buy a house, try to incorporate as many of these features as possible. It will be effort well spent.

# Ideas on Solar Home Heating

By William A. Shurcliff

**Definition of a Factor-of-Merit for Use in Comparing Designs of Simple, Two-Bedroom, Solar Heated Houses**

## Introduction

It is, of course, impossible to arrive at an objective, fair, accurate, generally acceptable scheme for **quantitatively** comparing various diverse designs of solar heated two-bedroom houses such as might be built in New England in 1974. Many different considerations—and different **kinds** of considerations—enter. Some are esthetic. Some are engineering. Different persons have different requirements. Some of the necessary engineering data are not available. Some schemes that may be practical in 1980 may be entirely impractical in 1974. Cost is crucial, yet prices of components are changing fast and unpredictably. The topographical setting of the house may have important bearing on choice of design.

Yet a designer cannot be content with such vague appraisals as "A is somewhat better than B." (Is it 10% better, or 1000% better? Would one or two minor improvements rescue B?). What is needed is a quantitative evaluation—even a very rough one.

It is intolerable, also, if the designer, each time he wishes to make quantitative comparisons of several designs, must think through afresh all the pertinent considerations and must afresh decide what weights to assign to the considerations. It is intolerable also that he has no ready way of explaining to colleagues what weights he used.

I here propose a factor-of-merit, hoping that friends will point out ways of improving it.

**Aspects excluded** To simplify matters I have ignored possible extensions of the solar heating system to: cooling the house in summer, supplying energy to the domestic hot water supply.

Also, I have ignored the cost (and nuisance) of the extra-high-quality thermal insulation required for the house. All solar-heated houses in New England require such insulation; the insulation is costly; but the cost is roughly the same (I assume) irrespective of the choice of solar heating scheme. (I **do** mean to take into account any very special thermal-insulation requirements of the solar heating system itself.)

I ignore also the merits of individual components, per se. Thus I do not attempt to appraise the collector per se, or the energy store per se, or the control system per se. What really counts is the system as a whole.

I ignore efficiency. The house occupant wants warmth, convenience, economy, reliability. If all these goals are achieved, he does not care whether they are achieved efficiently or inefficiently.

## Details

I call the overall factor-of-merit F ne, where **ne** stands for New England.

I define F ne as the product of many constituent factors $F_1$, $F_2$, etc. Thus:

$$F\,ne = F_1F_2F_3F_4\ldots$$

The constituent factors are arranged in groups. There are three groups:

**esthetics:** how the house looks from outside and inside; adequacy of daylight in rooms; adequancy of view outward from the rooms.
**performance:** ability to keep house warm enough on a typical day in winter; ability to keep house warm throughout a succession of overcast winter days; ability to keep house warm even if electric supply fails for a day; convenience of use; freedom from various hazards, dangers, worries.
**economics:** low construction cost; low operating cost; speed of construction; ease of repairing and modifying the system; avoidance of use of strategic materials.

I have tried to make the set of constituent considerations an orthogonal one, i.e., a clean one in which each major consideration squarely governs one factor and plays no role in any other factor.

The constituent factors are of normalized type. Specifically the value of each factor is approximately **unity** for a "typical, good design", and the values range from slightly above unity to a value well below unity.

Each factor is of affirmative type: higher value indicates a more desirable solar heating system, low value indicates a poor system. The value **zero** is assigned to a condition that would render the entire system completely unacceptable.

**Note concerning use of normalization:** Use of normalization (typical, good performance indicated by unity, and other values ranging somewhat above or below unity) has several important merits:

If the designer later introduces additional factors, they have little effect on the overall product (on F ne, that is) because multiplying a quantity by a number close to unity affects it only slightly. Thus F ne results obtained in the new way can still be compared, at least roughly, with results obtained in the old way.

Conversely, if the designer omits a few factors for any reason (for example, if data are lacking and he does not know how to arrive at an approximation) the effect on F ne is small.

The value of F ne itself is somewhere near unity, i.e., is of "managable" size.

### Esthetics: Constituent Factors

$F_1$
Outside appearance. Appearance of house as judged by a visitor approaching it. Appearance of south wall and other walls, and of roof. Extent to which trees and shrubs may be present, and freedom to choose types, heights, and locations of trees and shrubs.
1.3 excellent; solar heating system imposes no limitations
1.0 fairly good
0.5 very poor: external appearance of house is very poor; no trees or shrubs can be tolerated at south side

$F_2$
Inside appearance. Extent to which size of windows, and their locations, are normal, permitting normal amount of daylight to enter and permitting occupants to enjoy normal views of the environs.
1.3 excellent; the system imposes no limitations
1.0 fairly good
0.2 very poor; no daylight enters, and there is no view.

$F_3$
Other esthetic aspects. Freedom from noises from blowers, dampers, etc., from smells of chemicals, hot organic materials, fungi, etc., and from other esthetic nuisances.
1.2 excellent
1.0 fairly good
0 intolerably bad

## Performance: Constituent Factors

$F_{11}$
Room temperature. Adequacy of nominal temperature of rooms on typical day in January, assuming that there is no bunching of overcast days, no use of furnace or electric heater, no snow, no failure of electric supply.
1.2 excellent: 23°C (73.4°F)
1.0 fairly good: 18°C (64.4°F)
0.7 13°C
0.3 5°C
0.1 0°C

*Note: A solar heating system that cannot keep the rooms hotter than 5°C on a typical day in January may nevertheless be of considerable value inasmuch as (a) at such temperature, water in pipes does not freeze, (b) electrical heaters can easily bring the temperature in one or two rooms up to 18°C at little cost. (c) In warmer months (March, e.g.) the solar heating system by itself may be able to keep the roosm at 18°C.*

$F_{12}$
Holding time. Length of sequence of January days throughout which the nominal temperature of rooms (18°C) can be maintained. The length depends mainly on size of energy store and partly on system's ability to collect radiant energy even on overcast days.
1.4 excellent: 10 days
1.0 fairly good: 5 days
0.5 2 days
0.3 1 day

*Note: Even if system has a holding time of only 2 days in January, the system may be valuable inasmuch as supplementary heat can be provided–and in warmer months the holding time is much greater.*

$F_{13}$ (Re Snow, Etc.) Ability to perform well during or just after a heavy snow of sleet storm. Snow and sleet may blanket the collector window and auxiliary mirrors, and may be difficult to remove.
1.2 excellent: unaffected by snow and sleet (or removal is immediate and automatic)
1.0 fairly good
0.6 very poor: small amount of snow or sleet halts collection, and removal of snow or sleet is very difficult

$F_{14}$ (Re Collection with Elec. Off) Ability to collect energy even when electric supply is off. (It may be off because of failure of the electric utility's generator, because of breakage of the supply line, or because of rationing regulations or deliberate and extreme brown-out.)
1.2 excellent: collection is unaffected
1.0 fairly good
0.8 none: collection ceases until the supply is reinstated.

*Note: Here we assume that the year is 1978, the energy crisis is severe, and the electric supply will be off eight times per winter for eight daytime hours on each occasion. If the energy crises were much more severe than this, the value 0.8 should be replaced by a much smaller value.*

$F_{15}$
(Re Delivery with Elec. Off) Ability to deliver stored energy to the rooms even when electric supply is off for eight hours twice per month.
1.2 excellent: delivery is unaffected
1.0 fairly good
0.7 none: delivery ceases until supply is reinstated

*Note: same comment as for preceding factor.*

$F_{16}$
Extent to which the utilization (assignment) of space within the main storeys of the house is at the occupant's discretion
1.3 excellent: the space is fully at his discretion
1.0 fairly good
0.3 very poor: half of the space is preempted by solar heating system, and this half includes several key locations

$F_{17}$
Extent to which the utlization of space within the attic and basement is at occupant's discretion
1.2 excellent: the space is fully at his discretion
1.0 fairly good
0.7 very poor: great majority of such space is preempted by solar heating system

*Note: Preemption of attic and basement space is less serious than preemption of space in main storeys. Thus the extreme values proposed here are not very far from 1.0*

$F_{18}$
Degree to which the system operates without attention by the occupants
1.2 excellent: even if occupant is forgetful, or absent, system continues to perform well
1.0 fairly good
0.3 very poor: occupants must keep the system on their minds almost continually and must adjustment about five times per typical day.

$F_{19}$
Degree to which system itself escapes damage despite unusually extreme conditions of irradiation, outdoor temperature, indoor temperature, or amount of energy in store
1.2 excellent: virtually impossible that system will be damaged
1.0 fairly good
0.2 very poor: various quite possible conditions are likely to result in serious damage to system

*Note: Some kinds of storage materials or containers can be damaged by extreme cold. Some kinds of storage materials, containers, and insulating materials can be damaged by very high temperature.*

$F_{20}$
Extent to which health of occupants is independent of malfunction or overstressing of solar heating system
1.1 excellent: full independence
1.0 fairly good
0.2 very poor; various quite possible situations could result in serious harm to occupants

*Note: We include here hazards from fire–or release of toxic chemicals–attributable to the solar heating system.*

$F_{21}$
Extent to which the solar heating system is free of threats from external agents such as hail, rain, bees, wasps, birds, dogs, children, vandals, acid vapors in atmosphere.
1.2 excellent: system is fully free
1.0 fairly good
0.3 very poor: such agents might well do much harm

$F_{22}$
Extent to which the solar heating system is free of threats from internal agents such as rats, mice, centipedes, fleas, snakes, fungus, mushrooms, algae, children, dogs, cats.
1.2 excellent: system is fully free
1.0 fairly good
0.3 very poor: such agents might well do much harm

*Note: In some solar heating systems the energy store may, under some circumstances (especially in summer) become damp and thus may invite undesirable flora and fauna. Some systems employ non-enclosed running water. Some systems employ myriad small spaces for flow of air or water, and even small amounts of flora could fill–and block–such spaces.*

## Economics: Constituent Factors

$F_{31}$
Extent to which the cost of constructing the solar heating system can be kept low. (Here we **exclude** the cost of inventing the system and the cost of the engineering planning.) If expensive, specialized mass-assembly tools and assembly lines would be required (and would presumably be applicable to an undertaking to build **many** such heating systems) we assign to the individual house 5% of such large expense. We exclude also the cost of the land and costs associated with insuring that sunlight approaches the collector without obstruction. We exclude the cost of the very-high-performance thermal insulation required in house walls and roof (but we include, of course, the cost of special insulation that is part of the solar heating system). We exclude the cost of any **supplementary** heating system.
1.5 construction cost is less than 5% of cost of house as a whole
1.0 Construction cost is 10% of cost of house as a whole
0.7 construction cost is 15% of cost of house as a whole
0.4 construction cost is 22% of cost of house as a whole

$F_{32}$
*Extent to which annual cost of operation and normal servicing can be kept low. (Excluded are occupant's efforts and worry, major unexpected repair work, and cost of oil, electricity, etc. for supplementary heat.)*
*1.3 excellent: annual cost is extremely low, i.e., less than 0.1%* of cost of house.
1.0 fairly good, i.e., 0.3% of cost of house.
0.5 very poor, ile., 1.5% of cost of house.

$F_{33}$
Extent of freedom from long delays in construction as consequence of unique and difficult techniques involved
1.3 excellent: no significant

difficulties
1.0 fairly good
0.7 very poor: difficulties may cause 1 yr delay

$F_{34}$
Ease of making major repairs. Here we include especially great difficulties connected with repairs of components. Example: replacing broken window panes, including the least accessible panes; repairing or replacing tanks that have developed leaks as consequences of corrosion.
1.3 excellent: it is almost inconceivable that such difficulties will arise; or repair would be extremely simple
1.0 fairly good
0.5 very poor: such difficulties may well arise and repair would be very difficult

$F_{35}$
Extent of freedom from possibilities of serious shortcomings in design or construction. Here we take into account the dire consequences that sometimes result when daring new designs (never fully tested) are put into effect. Costs may soar. Performance may fall far short of expectation
1.3 excellent: design and construction are so straightforward and simple that it is almost inconceivable that serious shortcomings will appear
1.0 fairly good
0.7 very poor: shortcomings may well show up

$F_{35}$
Extent of avoidance of use of materials and skills that are in short supply. Here we take account of the national strategic importance (not dollar costs) of such materials and skills.
1.3 excellent: no strategic materials or skills required.
1.0 fairly good
0.5 very poor: much use is made of such materials and skills

$F_{37}$
Extent to which system layout, etc., could be changed, from time to time, as the owner's requirements change.
1.3 excellent: large changes could be made easily
1.0 fairly good
0.8 very poor: very difficult to change main features

$F_{41}$
All other considerations (numbers to be assigned as appropriate)

## ILLUSTRATIVE EXAMPLES

Table 3 shows some crudely arrived at results (estimates) for three solar heating systems:

**System S-20** It employs crude cylindrical horizontal mirrors that slightly focus radiation onto the underside of a thin horizontal tank the upper surface of which is insulated. The hot water flows to various storage tanks.

In proposing falues of $F_1$, $F_2$,. . . $F_{41}$, I have assigned low values to factors relating to holding time, snow, utilization of basement space, external threats, strategic materials—and to external appearance.

The product, i.e., F ne, is found to be 0.13.

This is a rather low figure. It disappoints me, inasmuch as I had thought some weeks ago that this system was excellent.

Table 2

Work Sheet for Computing Factor of Merit

| | Value Assigned | |
|---|---|---|
| | System | System |
| **Esthetics** | | |
| $F_1$ outside appearance | | |
| $F_2$ inside appearance | | |
| $F_3$ other esthetic aspects | | |
| **Performance** | | |
| $F_{11}$ room temp. achieved | | |
| $F_{12}$ holding time | | |
| $F_{13}$ re snow | | |
| $F_{14}$ re collection with elec. off | | |
| $F_{15}$ re delivery with elec. off | | |
| $F_{16}$ re space in main storeys | | |
| $F_{17}$ re space in basement & attic | | |
| $F_{18}$ re occupant attention needed | | |
| $F_{19}$ re proof against damage | | |
| $F_{20}$ safety of occupants | | |
| $F_{21}$ re external threats | | |
| $F_{22}$ re internal threats | | |
| **Economics** | | |
| $F_{31}$ construction cost | | |
| $F_{32}$ operating cost | | |
| $F_{33}$ delays in construction | | |
| $F_{34}$ ease of repair | | |
| $F_{35}$ re design errors | | |
| $F_{36}$ re strategic materials etc. | | |
| $F_{37}$ re ease of alteration | | |
| **Other** | | |
| $F_{41}$ all other | | |
| F ne (the overall product) | | |

**System Used in Certain New Solar Heated Houses in France** I have few facts on this system. I understand there is, at the south side of the house, a 1-m thick concrete wall which receives radiation passing through a window, and stores energy, and (via controlled circulation of air) delivers energy to the rooms.

I assume that the massive wall is ugly—from outside and inside; that the irradiated face of the wall becomes very hot and thus radiates much energy out through the window and accordingly collects only a modest fraction of the incident energy; that locating of view windows, etc., is greatly restricted; that construction cost is high (about $50 per m³ of concrete, plus other high costs); that the wall etc. are hard to alter. Accordingly I have assigned low values to the pertinent constituent factors.

I have assigned high values to most other factors. The overall products F ne is found to be 1.07—quite high.

**System S-58a.** Radiation transmitted by the collector window heates an upward stream of air, and the air then flows into the basement and passes through a high-pneumatic-conductance stack of perforated concrete blocks (32,000 kg of blocks; cost: $600.) A 1½ HP electric motor drives a blower that circulates the air. Air from the region surrounding the stack heats the rooms.

I have assigned low values of constituent factors pertaining to inability to collect energy when electricity is off, limited ability to heat rooms when electricity is off, and preemption of approximately half of the space in the basement. All other factor values are high.

The product, F ne, is 1.58, which I regard as attractively high.

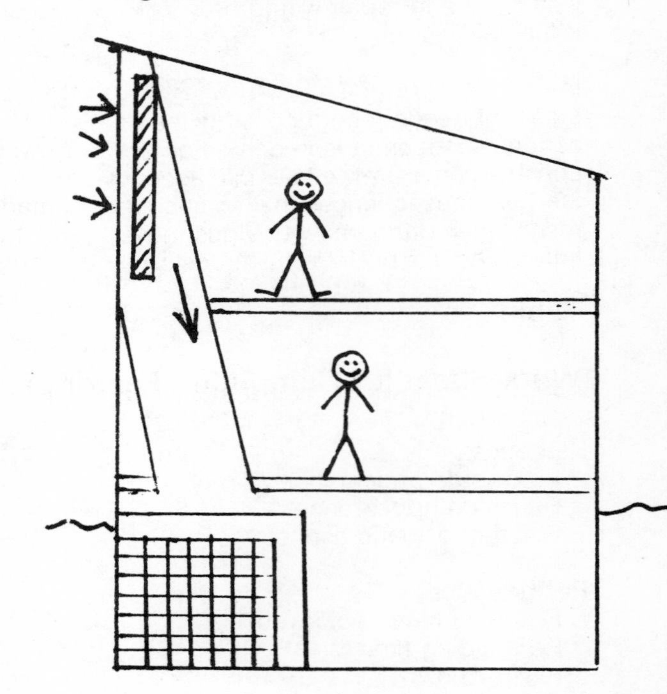

## Table 3
## Factor of Merit for Three Particular Systems

| | S-20 | S-58a | French |
|---|---|---|---|
| **Esthetics** | | | |
| $F_1$ outside appearance | | | |
| $F_2$ inside appearance | 0.8 | 1.0 | 0.7 |
| $F_3$ other esthetic aspects | 1.0 | 1.0 | 0.6 |
| **Performance** | | | |
| $F_{11}$ room temp. achieved | 1.0 | 1.0 | 0.7 |
| $F_{12}$ holding time | 0.8 | 1.2 | 0.9 |
| $F_{13}$ re snow | 0.8 | 1.2 | 1.2 |
| $F_{14}$ re collection with elec. off | 0.9 | 0.8 | 1.2 |
| $F_{15}$ re delivery with elec. off | 0.9 | 0.8 | 0.8 |
| $F_{16}$ re space in main storeys | 1.2 | 1.1 | 0.6 |
| $F_{17}$ re space in basement & attic | 0.8 | 0.7 | 1.2 |
| $F_{18}$ re occupant attention needed | 1.0 | 1.0 | 1.0 |
| $F_{19}$ re proof against damage | 0.9 | 1.2 | 1.2 |
| $F_{20}$ safety of occupants | 1.1 | 1.1 | 1.1 |
| $F_{21}$ re external threats | 0.7 | 1.2 | 1.0 |
| $F_{22}$ re internal threats | 1.2 | 1.0 | 1.0 |
| **Economics** | | | |
| $F_{31}$ construction cost | 0.7 | 1.0 | 0.6 |
| $F_{32}$ operating cost | 1.0 | 1.0 | 1.0 |
| $F_{33}$ delays in construction | 0.8 | 1.1 | 1.1 |
| $F_{34}$ ease of repair | 0.9 | 1.0 | 0.8 |
| $F_{35}$ re serious shortcomings | 0.9 | 1.0 | 1.0 |
| $F_{36}$ re stratetic materials etc. | 0.8 | 1.0 | 1.0 |
| $F_{37}$ re ease of alteration | 1.1 | 1.1 | 0.7 |
| **Other** | | | |
| $F_{41}$ all other | | | |
| F ne (the overall product) | 0.13 | 1.58 | 1.07 |

The choice of categories and choice of numbers assigned thereto are arbitrary and far from precise, and the overall values of F ne likewise are far from precise. Nevertheless it seems to me that the effort to list and evaluate the factors in this manner is helpful and the final comparison of F ne values is interesting.

## Crude Device for Solar-Heating a South Room of an Existing House: An External, Window-Mounted, Air-Type Collector (Type S-90)

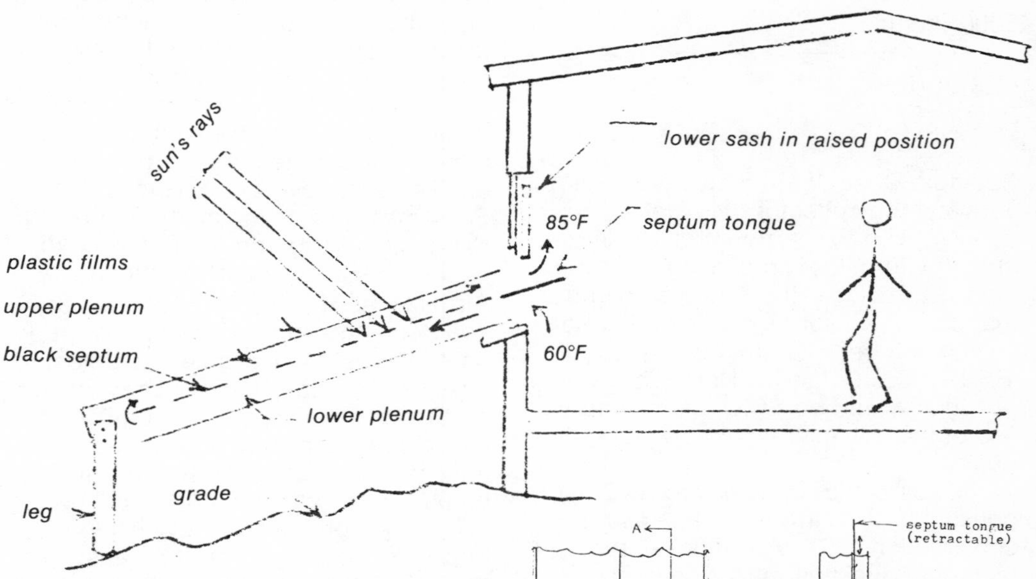

**Summary** Fig. 1 shows the general design.

At the start of a sunny day in January a resident opens the lower sash of the window to which the collector is attached and extends the collector's septum-tongue 1 ft into the room. Cold (~60°F) air from the room travels south and downward along the lower plenum, then travels north and upward along the upper (sunlit) plenum, then (at ~85°F) enters the room and rises toward the ceiling. The amount of energy delivered to the room is ~5000 Btu (~1.5 kWh) per hour, or ~25,000 Btu (~7.3 kWh) per day. Late in the afternoon the resident telescopes the septum-tongue and closes the sash.

Cost of plywood, insulation, plastic film, weatherstripping, etc.: $70 (guess).

Note that the collector operates automatically, by gravity convection. The flow-rate adjusts itself automatically to changes in level of irradiation. No electric power is needed; normal operation continues even during failure of electric supply. The dynamic thermal capacity of the collector is so low (~0.06 Btu/(ft$^2$,°F)) that delivery of hot air can start within 2 minutes after the sun comes out from behind heavy cloud cover. The resident can close the sash at any time. If heavy clouds obscure the sun for a half-hour, say, with sash left open, little room-energy is lost because (a) the sky itself delivers some radiation to the collector, (b) the collector is insulated on all sides, and (c) if the air in the collector becomes cold, the flow ceases (the cold air in the collector is trapped there by virtue of its greater density). The device can be transported from factory to home on car-top—or can be assembled at home by persons familiar with simple tools. It can be mounted in one hour by two such persons.

### Details

**Introduction** The design has been worked out at the request of W. J. Jones of the MIT Energy Lab., who pointed out the great need for a cheap, easily constructed device that can be "bolted onto" an existing standard-type house to provide a moderate amount of solar heat to at least one room—to save fuel, save money, and provide a little warmth even in the event of total failure of fuel supply or electric supply.

The design employs a simple but effective principle invented and tried out some years ago by S. Baer of Zomeworks Corp., PO Box 712, Albuquerque, N.M. 87103.



Fig. 1. The S-90 Collector

**Design** Figs. 2 and 3 show many details of the design. The collector is 4 ft x 8 ft in area and is 1 ft deep. The bottom, sides, septum, and spine are of ½-in. weatherproof plywood, and are secured by screws. Most surfaces are insulated with ½-in. polyurethane foam. The top consists of two, spaced, 0.004-in. films of weatherproof transparent plastic (Tedlar) secured at the edges by means of pressure-sensitive tape. Insulator surfaces that receive radiation are black; the large horizontal black surface has vestigal ribs that encourage creation of some turbulence in the air flowing along there.

An adapter-panel at the upper end of the collector is secured to the outside face of the window frame by means of screws, and is weather-stripped. The lower end of the collector is supported by two short legs (or two long poles, if device is to be mounted at a second-story window).

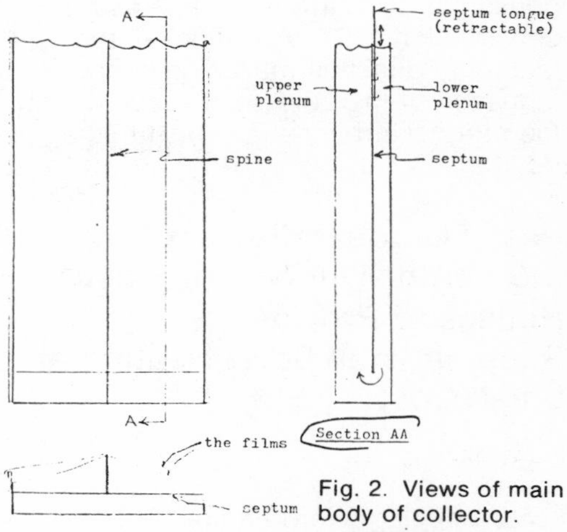

Fig. 2. Views of main body of collector. (Adapter-panel not shown; difficult to draw; details not worked out.)

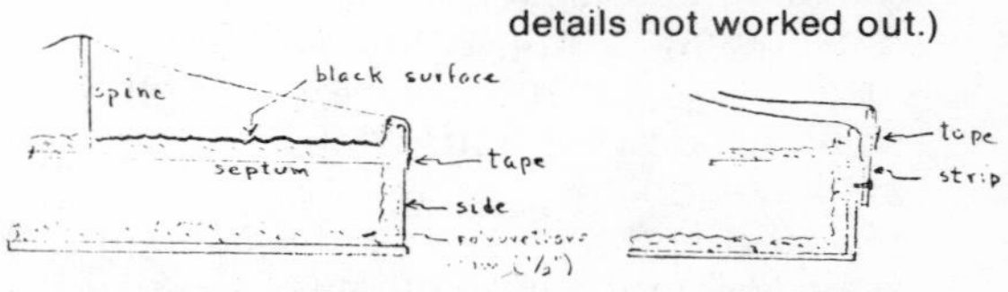

Fig. 3. Diagrams show locations of insulation. Also they show the means of affixing the plastic films: first film is taped to collector side; then auxiliary strip is screwed on and second film is taped to it. The scheme provides an airspace between the two films. The spine increases cross-section of upper plenum, keeps films taut, and helps "slope" them to shed rain etc.

**Performance** The nominal area of the collector is 4 ft x 8 ft = 32 ft$^2$; the effective area is 30 ft$^2$. On a sunny day in January the amount of solar radiation received in (~200 Btu/ft$^2$) (30 ft$^2$) = 6000 Btu, of which ~5000 Btu (~1.5 kWh) is delivered to the room. The amount delivered during the day as a whole is (~5 hr) (5000 Btu/hr) = 25,000 Btu (~7.3 kWh). Reference to my report S-58 of 6/19/73 shows that this amount of energy is ~4% of the total energy requirement of a well-insulated two-bedroom house in Boston on typical 24-hr day in January, and is ~7% of the amount of energy required of the furnace (much energy being provided also by miscellaneous sources such as stove, lights, appliances, human bodies, sunlight entering windows).*

** On a sunny day, the energy delivered by the collector to the room in question would keep this room much warmer than other rooms–75 to 80 °F as compared to 60°F, say. Any occupant who feels chilly can come to this room to warm up.*

**Temperature rise** If the linear rate of air-flow is 1.5 ft/sec (guess) and the plenum cross-section is 2 ft², the volume flow is (1.5 ft/sec) (2 ft²) = 3 ft³/sec, the mass-flow rate is (3 ft³/sec) (0.08 lb/ft³ = 0.24 lb/sec, and accordingly the temperature rise is:

$$\frac{5000 \text{ Btu/hr}}{(0.24 \text{ lb/sec})\ (3600 \text{ sec/hr})\ (0.24 \text{ Btu/(lb,°F)})} = 24\ °\text{F}$$

**Savings. Cost.** If, on a sunny day in January, the collector cuts fuel consumption by 7%, it will cut it by larger amounts in milder months. On overcast days the device will not be used. I guess that, for the winter as a whole, the fuel saving will be 15%. If, in the winter of 1974 - 1975, the fuel cost for the house in question would be $600, the saving for use of the collector would be (15%) ($600) = $90.

The cost of the collector may be $170, I guess; i.e., $70 for materials, $50 for construction, $50 for selling—if the factory in question makes several thousand of the devices. I assume that the purchaser himself will install the device.

Thus in terms of reduction in amount paid for fuel, the device would pay for itself within 3 years. In terms of reduction in worry concerning possible failure of fuel supply or electric supply, it might justify itself in an even shorter period.

**Drawbacks** Shrubs may be in the way of the device (but it may happen that the collector harmlessly "reaches over" the shrubs). Trees may shade the collector (but deciduous trees will produce little shade). Children, cats, squirrels, birds may walk on the plastic films and damage them (but fences or chicken-wire covers could prevent this). Snow may stick on top of the device (but residents can scrape it off, and in any event the sun will soon melt it off). Air-leaks at window may occur unless adapter is secured and weather-stripped carefully (but the leaks can be detected easily and plugged easily). The warm air produced is delivered to the ceiling (but this increases the amount of radiant heat from the ceiling; also, an electric fan can be used to circulate air from ceiling to floor). In a high wind the device might be ripped loose (but it would be easy to secure the legs to stakes driven into the ground, or to a very heavy log). The expected life of the plastic is only a few (5?) years (but replacing it would be simple).

**Some Options** Install such devices on several 1st & 2nd story windows.

Employ a small electric fan to speed the flow of air along the collector and thus increase the amount of energy collected (by 10%?).

If a fan is employed, the device may be mounted horizontally; no slope is needed.

Employ a special adapter which would permit mounting the device **vertically** below a second-story window.

On the two sides of a sloping collector, install flaring aluminized "wings", to funnel additional amounts of radiation into the collector. Tie 12-in. diameter bundles of brush to the upper edges of the wings to dampen wind-gusts and reduce average wind-speed past collector top by a large factor (factor of 3?). Thus reduce losses; increase the efficiency of collection (by ~7%?).

Provide an alternative design that is longer and wider; use fan.

Provide an alternative design that runs parallel to the house (and uses fan, and does or does not slope). Such device can be very long. It benefits from radiation reflected from white side of house.

## Use of Crude Reflectors Outside the South Windows of Existing Houses to Provide Supplemental Solar Heating at Small Cost

### Summary


A surprisingly large amount of supplementary solar heating of south rooms of an existing house in sunny location can be provided by installing—just outside the sills of the south windows—crude, flat, slightly-sloping, plywood-and-aluminum-foil reflectors. Using several such reflectors, costing ~$200 in all, the owner can not only save ~$50 to $100 of purchased energy each winter but also enjoy extra warmth in those rooms and reduce his worries concerning fuel shortages and electrical blackouts.

The proposed reflectors are cheap, durable, easily installed. Their effective thermal capacity is zero: delivery of energy to the rooms starts instantly when the sunlight strikes the reflectors. The reflectors make **no** contribution to heat losses at night, etc. In summer the reflectors can be re-mounted so as to serve as blinds or awnings to exclude radiation and keep the rooms cool.


### Details

Fig. 1 shows a conventional house with four types of reflectors mounted adjacent to south windows (here assumed 4 ft wide and 5 ft high). Fig. 2 shows the (5°) tilt of the reflector. Fig. 3 shows alternative shapes and sizes of reflector—to serve one window or an adjacent pair of windows.

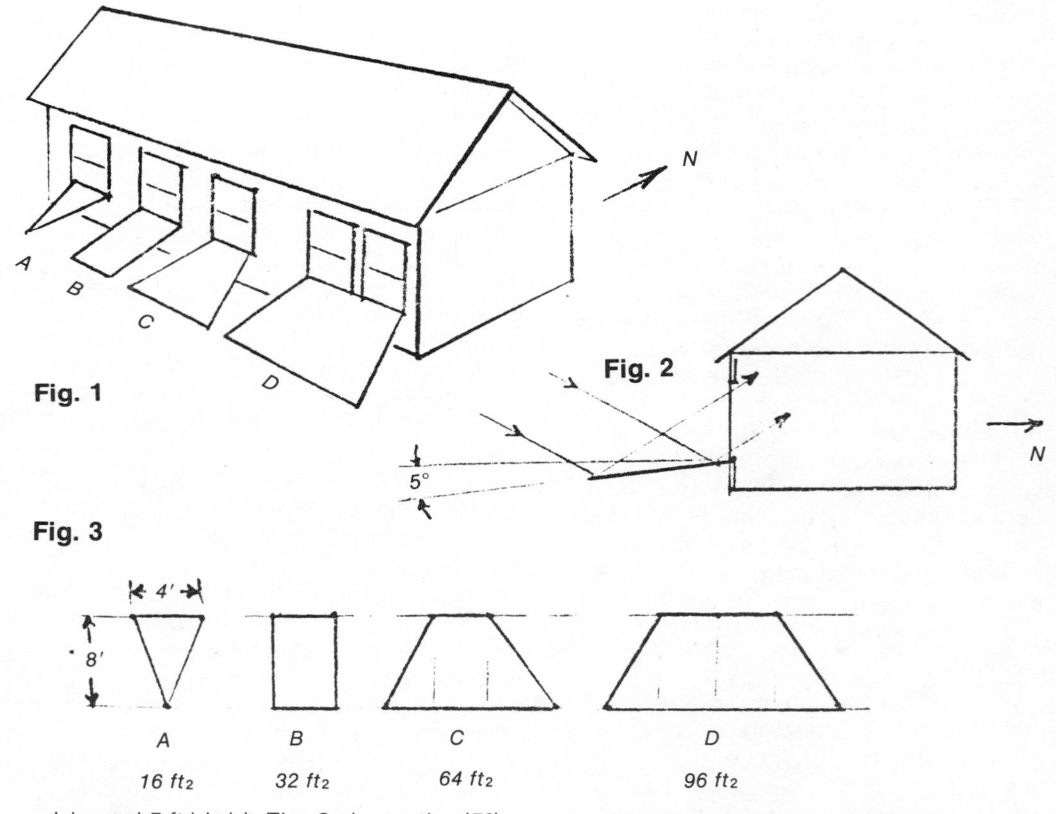


**Tilt of Reflector** For houses at latitudes such as that of Boston (42 °N), the altitude of the sun at noon on a typical day in midwinter is, say, 30°, and at noon on such day a Type B reflector (4 ft x 8 ft) that is tilted 1 or 2 degrees downward-toward-the-south nicely fills the window with reflected sunlight. But an elevation angle of 25° is more nearly typical of a sunny hour on such day, and for this elevation angle a reflector tilt of 5° is about optimum.

**The Reflecting Surface** The reflector might consist of ⅜-inch sheet of weatherproof plywood to which sheets of shiny aluminum foil (or aluminized mylar) have been affixed—by glue, tape, or other means. (The owner could install a fresh reflecting sheet every few years, if necessary.) Or one could use plywood to which a thin aluminum sheet had been bonded at the factory. (Some benefit would be achieved using, merely, aluminum paint or white paint.)

**Relative Efficiencies of the Four Types of Reflector** In terms of amount of energy delivered per unit area of reflector, the Type A device is better than B or C. All parts of it are so close to the window and/or so close to the north-south centerline of the reflector that they deliver energy to the window throughout a considerable portion of the day. Type B, although delivering more energy, is slightly less efficient inasmuch as the outer corner parts are effective throughout a slightly shorter portion of the day. Type C, although delivering even more energy, has even lower efficiency because some parts are even farther from the window and from the centerline and hence are effective for brief portions of the day only (and these portions occur when the sun is so low in the sky that the effective projected area of the reflector is relatively small).

Where there are two windows close together, Type D can be used especially effectively. Nearly every part of it delivers energy to one window in one portion of the day and to the other window in another portion of the day.

I estimate the efficiencies of Types A, B, and C to be in the proportion 1.1, 1.0, and 0.85.

To compute the relative efficiency of D compared to C, I note that D can deliver, through **one** of the windows of the pair, slightly more energy—say 15% more—than C can deliver, and I also note that the area of D is 150% that of C, so that the Reflector-D-area **per window** is only 75% the area of C. Thus the relative efficiency of D compared to C is (1.15) (1/0.75) = 1.5. The relative efficiency compared to B is (1.5) (0.85) = 1.3.

**Direct Energy vs. Reflected Energy** It is worth noting that the amount of energy entering the windows directly from sun and sky is greater than the amount entering via the reflector—because of the reflector's imperfect reflectance, imperfect tilt, limited angular subtense.

**Curvature of Reflector** It would be slightly beneficial to bend the reflector slightly (make it slightly concave upward) and to allow the south corner portions to droop slightly (to compensate for the low elevation of the sun at times (remote from noon) when these portions come into play). But such changes would probably not be worth the effort.

**Amount of Energy Delivered by the Reflector**

**Type B** Near noon on a sunny day in January with the sun at 25° above the horizon, the amount of radiant energy proceeding directly from sun to the Type B reflector in an hour is
(200 Btu/ft$^2$) (32 ft$^2$) (sin(25° + 5°)) = (200) (32) (0.5) Btu = 3200 Btu.
Of this, ~75% is specularly or near-specularly reflected, ~95% of this encounters window glass (rather than muntins), and ~84% of **this** is transmitted by the (double-glazed) window. Thus the net amount of energy delivered to the room is (3200 Btu) (0.75) (0.95) (0.84) = 1900 Btu. A sunny January day as a whole has direct solar radiation corresponding to about 5 hours of noon radiation, but to allow for the unfavorable vertical and horizontal angles at times well before noon and well after noon, I use the figure 3½ hr, rather than 5 hr. Thus the amount of energy delivered on a January sunny day as a whole is (3½) (1900 Btu) ≅ 6700 Btu. Some contribution will be made by diffuse radiation; that is, the sky contributes more energy (via reflection from reflector) than would be contributed (in the absence of the reflector) by the ground beneath the reflector; I guess the contribution from diffuse sky radiation to be 10% of the contribution by direct solar radiation, i.e., (0.10) (6700 Btu) = 670 Btu. Accordingly the sunny-January-day total energy delivered is 6700 + 670 ≅ 7300 Btu. To allow for the fact that a typical day in winter has a longer period of sunlight than a January day has, I add 10%, to obtain ~8000 Btu per typical sunny day in winter.

**The Four Types** Taking into account the relative areas of the reflectors (see Fig. 3) and the relative efficiencies (see a previous section), I estimate that the amounts of energy delivered to the room on a sunny day in winter to be:

| | | | |
|---|---|---|---|
| A: | 1.1 (16/32) (8000 Btu) | ≅ | 4,000 Btu (~1 kWh) |
| B: | 1.0 (32/32) (8000 Btu) | ≅ | 8,000 Btu (~2½ kWh) |
| C: | 0.85 (64/32) (8000 Btu) | ≅ | 14,000 Btu (~4 kWh) |
| D: | 1.3 (96/32) (8000 Btu) | ≅ | 33,000 Btu (~10 kWh) |

**Value of Energy Delivered** If the energy delivered by the Type D reflector were provided, instead, by electric heaters consuming electric power costing 4¢/kWh, the cost of such power would be (10 kWh) ($0.04/kWh) = $0.40 per day. If the reflector were used throughout the equivalent of 120 sunny days per winter, the energy supplied would be worth (120 day) ($0.40/day) = $48.

I estimate, similarly, that for A, B, C, and D the amounts of energy provided per winter would be worth (in terms of electric power cost) $6, $12, $20, and $48 respectively. Relative to energy obtained from burning oil—at tomorrow's price of oil—the amounts of energy in question would be worth about half as much.

**Cost of Reflector** The Type B reflector, consisting of a 4 ft x 8 ft. panel of ⅜-inch weatherproof plywood (costing $10) to which an aluminum coating (costing, say, $3) has been affixed, would have a total cost of $25, I estimate. Costs of Reflectors A, B, C, D would be about $15, $25, $45, $65.

**Money Saving** Reflector D would "pay for itself" in 1½ years, relative to electric power, and in 3 years relative to use of oil. It would take the other types of reflector about half again as long to pay for themselves.

**Other Benefits** The reflectors would give the residents satisfaction from (1) having one room that is ~10°F (say) hotter than typical rooms (i.e., 70°F instead of 60°F, say) on sunny days in winter and (2) being somewhat less dependent on continuity of oil supply and continuity of electric supply (essential for operation of an oil furnace). If several reflectors were used, the residents would have the satisfaction of knowing that, even if, in midwinter, the fuel supply or electric supply failed during a period in which there was a typical amount of sunny weather, the house would keep above 32°F and accordingly no water pipes would freeze and burst.

**Note**: On a sunny day in winter, seven Type C devices—or three Type D devices—would supply about 100,000 Btu, which is about ⅓ of the amount of energy the furnace would normally be called on to supply on such day—for a typical, well-insulated two-bedroom house near Boston (per my Report S-58).

**Desirability of Installing Additional Windows** A homeowner might find it desirable to install additional (double-glazed) windows beside "lone" windows, or between a pair of windows that are 5 ft. apart—to obtain more radiation directly through such windows and to permit installing Type D devices to contribute further to the partial solar heating of the house.

**Installing the Reflector** The reflector-edge closest to the house could be screwed, nailed, or tied to the window will, or to some wooden blocks previously screwed to the side of the house. The outer edge of the reflector could be fastened to stakes or posts, or could be rested on (and tied down to) a fence, wall, or heavy horizontal log. The reflector should be sufficiently secure to be unaffected by winds.

**Summer Use of Reflector** Instead of

storing the reflector in a garage in summer, the homeowner might raise up the outer end so as to form a blind for the window. Or he could raise both ends so that the device would form an awning, or sunshade. Thus the room could be kept relatively cool even on a sunny hot day in summer.

**Adjusting Tilt of Reflector** The residents might find it worthwhile to adjust the tilt of the reflector slightly, from month to month, to increase its efficiency. In December, e.g., greater tilt is desirable.

If the "south" side of the house in fact faces 20° west of south, the residents might find it worthwhile to install the reflector so that it has a slight (~10°) tilt upward-toward-the-west.

**Installing Reflectors on Second-Story Windows** This can work out very well. Here, the reflectors will not interfere with shrubs or walk-ways, and they are less prone to the shaded by nearby trees. The reflectors may be made shorter to avoid unduly shading windows below. The outer end of the reflector can be supported by poles or by long brackets, or by tie-rods running from the overhead eaves.**Acutal Sizes of Windows** Most windows of most houses are much smaller than the size (4 ft x 5 ft) assumed in the present calculations. Thus smaller reflectors would be used—and, preferably, more of them. Being smaller, they would be easier to handle and install, but the overall cost per unit area would be somewhat greater.

## Discussion

The system is cheap, simple, rugged, has no moving parts, requires no adjustments, uses no electricity. It has zero effective thermal capacity: it starts delivering energy to the room the instant sunlight strikes the reflector. Even on somewhat overcast days it delivers some energy to the room.

The system makes **no** contribution to heat losses. It adds nothing to the "shell" area of the house. It makes no change in the shell. At best, it adds much energy to the house. At worst (heavy overcast; or nighttime) it takes away no energy.

It somewhat disfigures the house-exterior, and the grounds adjacent to the south side of the house. The owner may have to remove snow from the reflector several times a winter (although snow itself is a fairly good reflector).

**Distribution and Storage Question** The reflectors help heat the south rooms, but not the north rooms—unless fans are used to move air from the former to the latter, or unless the furnace-heat distribution system can be adjusted to compensate for the unequal solar heating. Often, it may be feasible simply to let the north rooms (usually less important than south rooms) be a little cold!

Could some crude form of storage be provided, so that energy taken in on sunny and warm days could be stored for use on cold sunless days? Many crude schemes come to mind; they involve large volumes of water (in tanks, or in plastic or glass bottles, etc.), or large volumes of concrete (concrete floor, concrete ceiling, concrete wall—or channeled concrete blocks through which air is driven by fans or blowers. An elegant scheme would be to use a local stack of spaced trays containing eutectic salts, and to cause air to pass through the stack—all as arranged by Telkes et al in the Solar One House of the Univ. of Delaware. Another elegant scheme would be to use a small, high-COP heat-pump to take the energy from the air near the upper part of the room and deliver the energy to the conveniently located tank of water. Such schemes would be facilitated if the region just inside the window were largely enclosed, or isolated, by means of curtains or semi-transparent panels—to keep the heat concentrated prior to delivery to the store (but without blocking all light from the window and without blocking all view through the window).

**Grand Combination System** It might be especially efficient to combine the general scheme discussed above (providing, say, 35% of the winter's heat requirement) with a "standard" solar collection and storage system (providing 65% of the requirement). Thus 100% solar heating could be achieved relatively economically.

## Sun-Scoops to Be Installed at Windows at East and West Ends of House, to Produce Some Solar Heating

At several windows at east end of house, and several windows at west end of house, install sun-scoops consisting of flat, aluminized sheets of plywood. As indicated in the following sketches, each sheet is vertical and makes an angle of 30° with vertical east-west plane. Width of sheet is same as width of window. Height of sheet is same as height of window. Midpoint of sheet is higher than midpoint of window by about 1/5 of window-height.

Scoops at east end of house work fairly well from 9:30 to 12:15, and scoops at west end work from 11:45 to 2:30 (solar time). Amount of energy delivered to rooms: About the same as from the south-window-reflector devices described in great detail in my 12/27/73 report. Cost, too, is about the same.

Devices are cheap, rugged. Delivery of energy to the rooms starts instantly when the sunlight strikes the mirrors (scoops). The scoops cause **no** increase in heat-loss at night etc. They can be used in summer as blinds, to exclude sunlight.

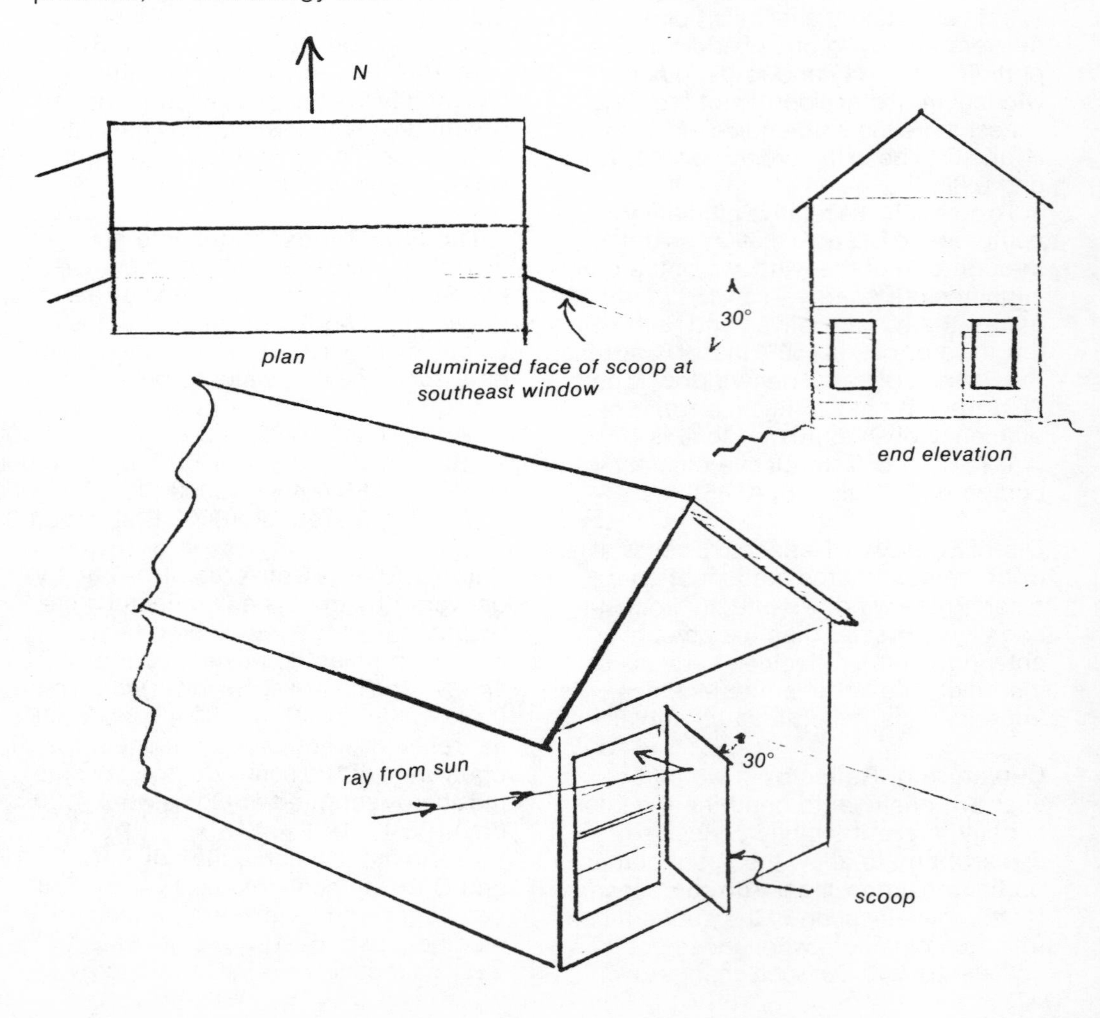

# Parabolic, Spherical, and Other Reflecting Surfaces

By Paul D. Ackley

With flat plate collectors, the maximum amount of heat collected is proportional to the blackness of the surface, but even with the finest of black surfaces there is a limit to the temperatures that can be obtained. The only way to get higher temperatures is to bend the light so that a larger amount of light will fall on a given area. There are two methods that can be employed to achieve more light per given area:

First, one could use a lens. This is similar to using a magnifier to burn a hole in a piece of paper with the light of the sun. Normally the light of the sun cannot heat the paper to combustion, but when concentrated by the lens a small portion of the paper can be heated to combustion. Lenses may be practical if one intends to burn holes in paper, but if a larger volume or higher temperatures are needed the lens becomes very expensive. Even fresnel lenses in sizes much larger than 20″ in diameter are expensive.

The second alternative open, within reason, is that of using reflective surfaces. Just as the convex surface lens focuses light to a point, the concave mirror does likewise. Unlike the lens, the reflector can be easily made at home out of a wide variety of materials, and characteristics of these reflectors are easily predicted with simple formulas which will be set forth below.

A little background information concerning reflective surfaces is necessary. The basic rule which must be stressed is: The angle of incidence is equal to the angle of reflection (see figure 1).

Using this rule and a little math, one can compute the way light bounces off any surface. Now all that needs to be done is to find out what shape of mirror would be most practical for collection of solar energy for a given purpose. The purposes seem to fall into two main categories: either low heat/high volume, or high heat/low volume. Low heat/high volume would be used to gather heat to run engines or heat a home, whereas high heat/low volume is used primarily to gain as high temperatures as possible, and is usually referred to as a solar furance. 60″ diameter parabolic mirrors used as solar furnaces have attained temperatures in excess of 7200°F.

The biggest advantage of the parabolic mirror over spherical mirrors is illustrated in figures 2 and 3. With the parabolic mirror, the parallel bundles of light from the sun are focused to a single point where the sun's rays strike the parabola perpendicular to a line tangent to its minimum point (the center). This gives good, even distribution of the light across a round tube placed at the focal point, or with a flat disc or strip placed at the focal point (see figure 4).

On the other hand, the spherical mirror does not focus the light to a single point. Only the light that strikes the very center region of the spherical mirror is reflected near the intended focal point. Both mirror diagrams are the same size (4″ across and 1″ deep) yet, as is easily visible, the area of high light concentration in the spherical mirror is large and uneven; that of the parabolic mirror is small and even. The more lines in a given area the higher the concentration of light.

The drawback of the reflector collector is that it has to be aimed at the sun and on cloudy days there is no sun at which to aim the reflector. The accuracy of this aiming is directly dependent on how

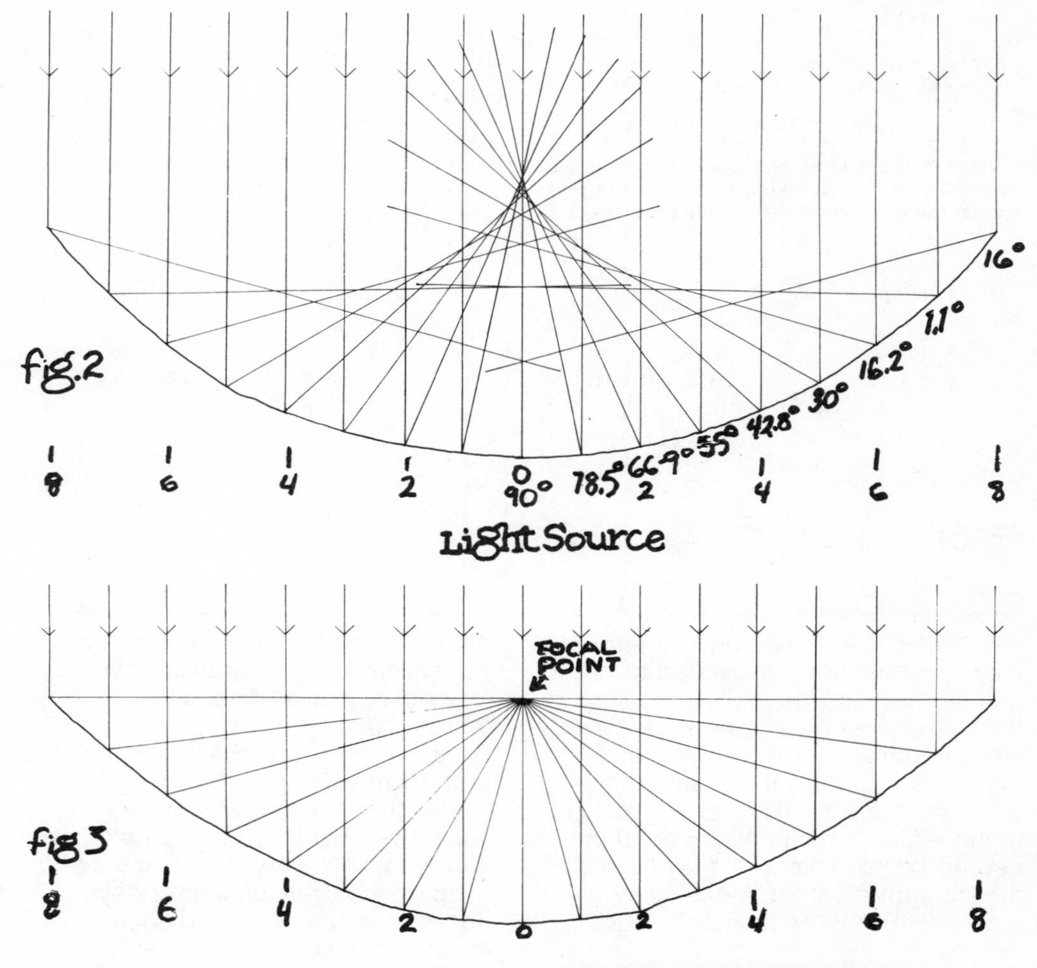


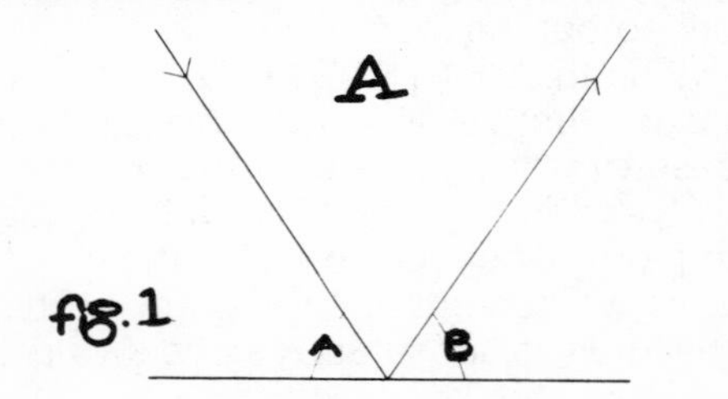


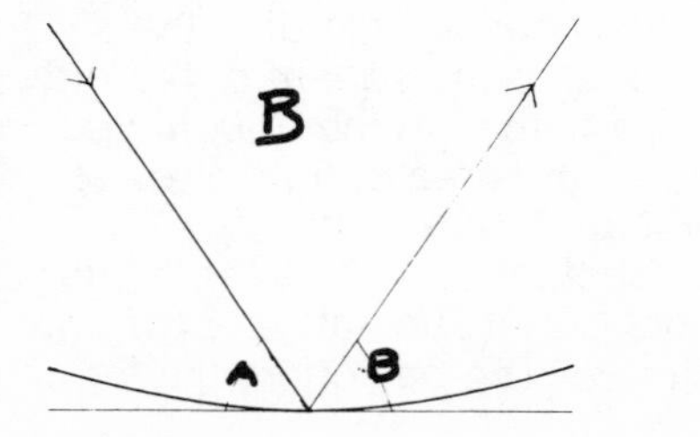


In the first example (on the left) the light strikes the surface at angle A and then is reflected off the surface. If angle B is measured it will always equal angle A. In the second example, the surface is curved, but because all light rays are so very small, they treat each point of the surface as a separate flat surface. A representation of this is drawn beneath the curved surface. The flat surface is tangent to the curve, which means it touches the curve at that point and only that point.

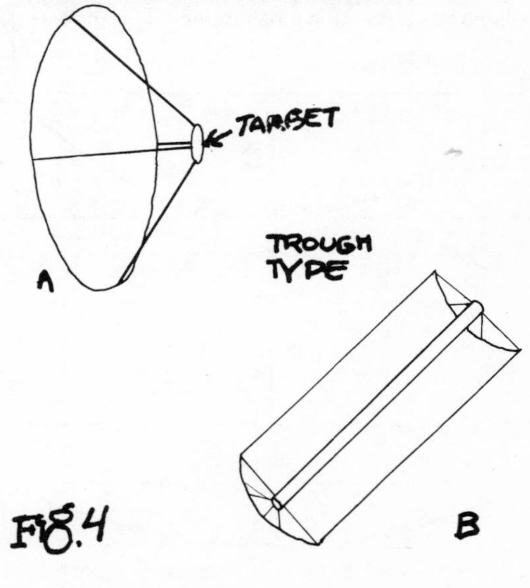

much one concentrates the light. Figure 5 shows with the 8″ reflector and ¾″ pipe setup that 12.5% of the light is lost when the suns rays are off just 6° on either side of the perpendicular. This loss is relatively small. If the angle of error off the perpendicular is increased to 12°, it can be seen that the reflected rays miss the pipe entirely (see figure 6). However, when the diameter of the pipe is enlarged to 2⅔″ (⅓ the width of the reflector), there is only a 19% loss of light in the region 30° on either side of the perpendicular. This would take in 60° of the sky with a probable efficiency in excess of 80%. The amount of light gathered beyond 30° soon tapers off to nothing as the light can no longer strike the pipe, but this ratio would provide a great deal more energy on a cloudy day than the same size collector with a small pipe. The only reason a small pipe might be used instead of a large pipe, would be to attain higher temperatures or to try to further reduce the cost of the solar collector, or to heat the contents of the pipe more evenly.

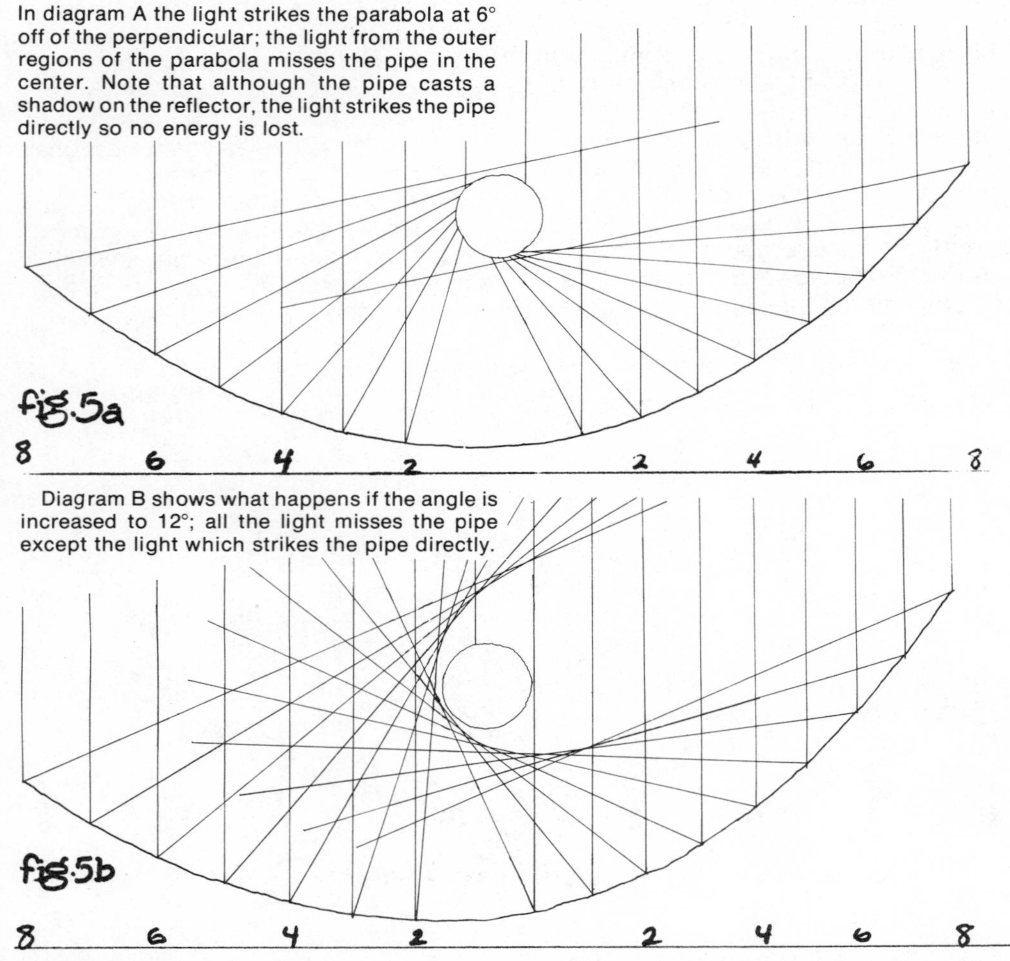


In diagram A the light strikes the parabola at 6° off of the perpendicular; the light from the outer regions of the parabola misses the pipe in the center. Note that although the pipe casts a shadow on the reflector, the light strikes the pipe directly so no energy is lost.

Diagram B shows what happens if the angle is increased to 12°; all the light misses the pipe except the light which strikes the pipe directly.

With the round solar furances it is imperative that they be accurately aligned with the sun and that they be used on relatively cloudless days because with these round solar collectors, a very slight misalignment can cause the sun to burn a hole in one of its mounts or it may not focus the light to an exact pinpoint, which would cause the maximum temperature of the furnace to be lowered. When compared with the trough-shaped reflector the round collector has a greater light loss from its outer regions when misaligned than the trough type. For example, half the light of an 8″ round collector comes from the outer 2 11/32″ of the mirror. This is true because the area of light which these round mirrors collect is directly proportional to the square of the radius times—, whereas in the trough-type collectors, the area of light is directly proportional to the width.

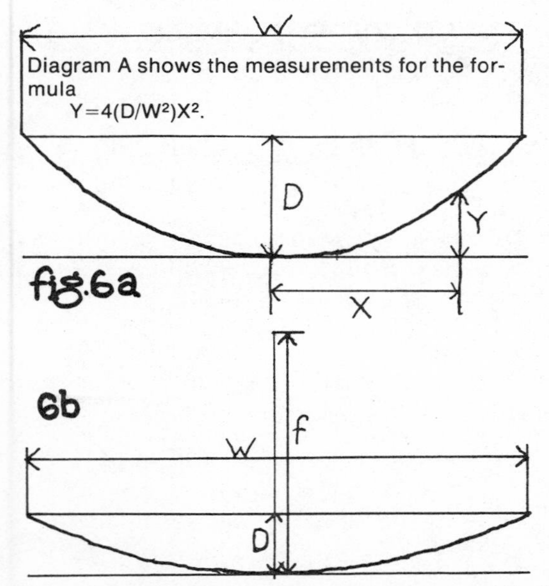


Diagram A shows the measurements for the formula $Y=4(D/W^2)X^2$.

Diagram B shows the measurements for the formula $f=(1/16)(W^2/D)$.

A parabola may at first seem hard to construct but mathematically they are very simple. The basic formula for a parabola is $Y=aX^2+bX+c$. However, the bX and the c terms have no real effect on the shape of the parabola; what they do is move the parabola around on the graph. Omitting bX and c, one is left with $Y=aX^2$, where X is the distance along the line tangent to the minimum point, from the point where it is tangent, and Y is the height under the parabola at each point X on this line. The 'a' is a constant for the equation which determines the actual size of the parabola. When this number "a" is large, the parabola which it defines is tall and skinny; when this number is small, the parabola which it defines is short and wide. In a more practical form, this "a" term is equal to $4D/W^2$, where "W" is the width of the parabola and D is the depth. Putting this new form of "a" into the original equation, one obtains $Y=(4D/W^2)X^2$.

Another useful item is locating the focal point. The focal point is difficult to find by working the math through; but the formula that derives from the math is pretty simple: It is $f=(1/16)(W^2/D)$.

The formula $Y=(4D/W^2)X^2$ gives you only the height under the parabola and requires that you know both the width and depth. The formula below may be more complex in structure but it would be more useful when trying to figure out your own parabolas. It is $D=(1/16)(W^2/f)-(X^2/4f)$ and it requires that you know the focal length (f) and the width.

Another thing one might want to know about a collector is how much light or power it will collect. The following formulae are based on the solar constant 1.92 small calories per square centimeter per minute; and allowing for a 19% loss of light due to the atmosphere. (On cloudy days more light is lost, sometimes in excess of 35% in addition.)

| | round type collector | trough type collector |
|---|---|---|
| amount of horsepower | .424 $R^2$ | .135 LW |
| number of BTU per hr. | 108 $R^2$ | 34.4 LW |

(Where R is the radius in feet, L is the length in feet, and W is the width in feet.)

The following is an example of how to use the formula:

$D= (1/16)(W^2/f) - (1/4f)(X^2)$. First assume that the collector that is needed should be 8″ wide and have a focal length of 4″. This means that W is substituted by 8 and f is substituted by 4. Substituting these values, the equation looks like:

$D= (1/16)(8^2/4) - (1/4\times4)(X^2)$.

Working out the math simplifies the equation to: $D=1-(1/16)X^2$. To plot this equation first set up a work line and measure off on either side of the center 4″ (half the width); divide this up into inches (the more divisions the better the accuracy); see figure 7, diagram A. The next step is to substitute X to the X value, which is how far the point is from the center; for example, in the center X=O, solve the equation for D and measure down that many inches. If this is repeated for all the points on the work line you will end up with something that looks like figure 7, diagram B. The final step is to connect the points with a smooth continuous line (use a French curve if you have one) and you will have the parabola as shown in figure 7, diagram C.

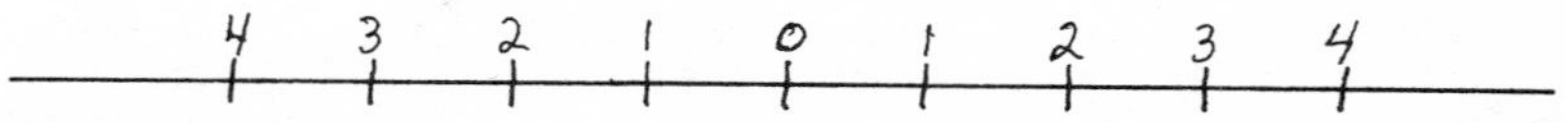

# RING PARABOLOID REFLECTOR

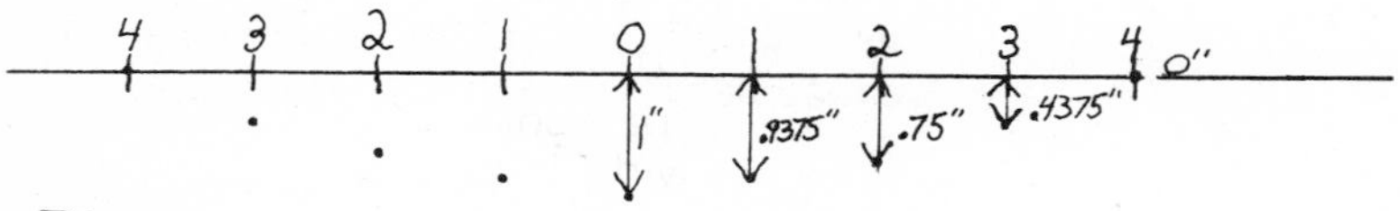

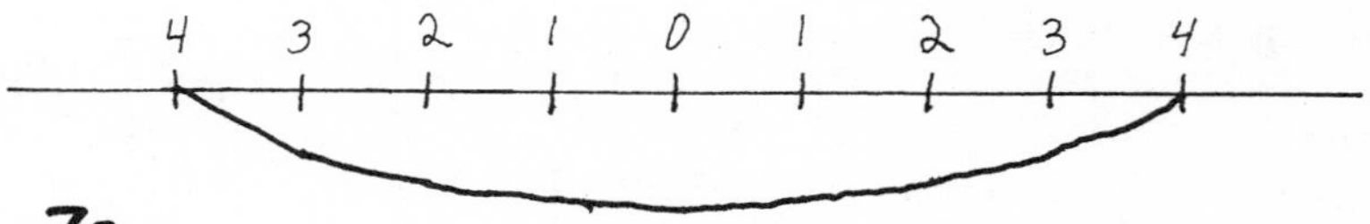

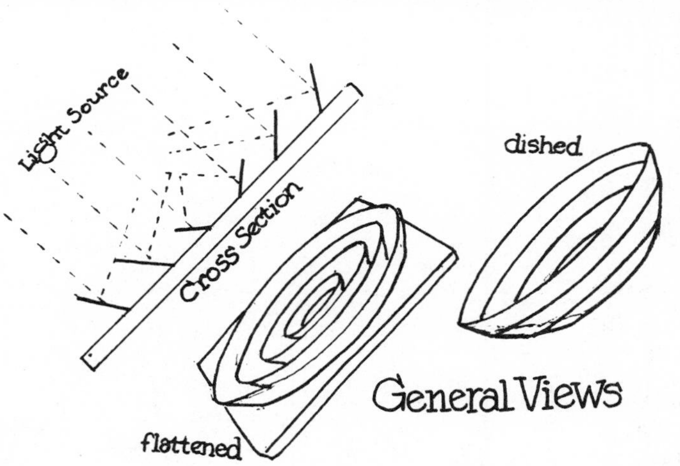

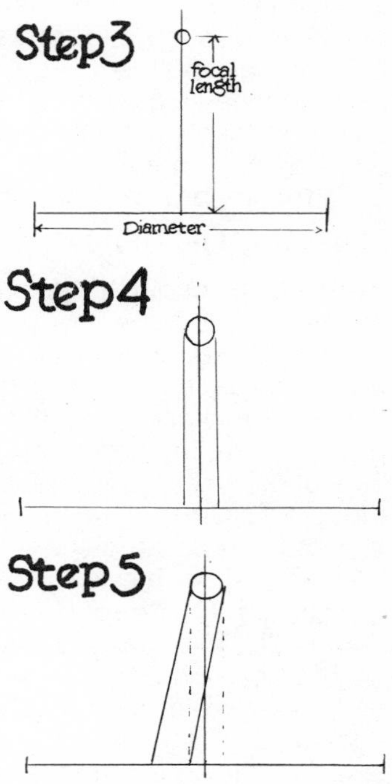

It isn't easy to make a parabolic reflector for high temperature solar experiments. It's much easier to approximate a paraboloid as a series of flat segments. A reflector made in this way cannot focus the sun's rays into a single point, but this is not always necessary or desirable. If the focus is too sharp, you run a risk of melting a hole in the bottom of your boiler or crucible or whatever instead of heating the whole thing evenly. This article describes how to make a reflector as a series of rings (actually conic segments). An advantage of doing it this way is that the whole reflector can be "flattened out". In this form, the reflector take up less space and is easier to handle.

Step 1: First decide on the diameter and focal length needed. A reflector with a long focal length will be shallower and easier to make than one with a short focal length.

Step 2: Decide on the size of the target the energy will be focused on. This determines the width of the rings. The narrower and more numerous the rings, the smaller the spot that receives the energy.

Step 3: Draw a centerline and baseline and indicate the desired focal length and diameter. Draw in the target spot at the focus.

Step 4: Draw a pair of parallel lines from the target spot to the baseline. A circle the same diameter as the focus spot will be the base of the reflector.

Step 5: Draw a pair of parallel lines from the focus spot that fall just outside the first circle (or the previous ring). The next segment must be angled so that the incident and reflected rays form equal angles. An easy way to determine this is to place a protractor at their apex and then adjust it until the 90° mark bisects the angle. (The incident ray is represented by a line perpendicular to the baseline that just touches the outside of the previous ring.) A line drawn along the protractor's base then gives the correct angle for the new segment.

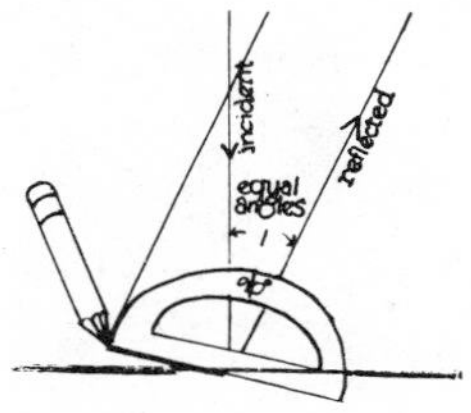

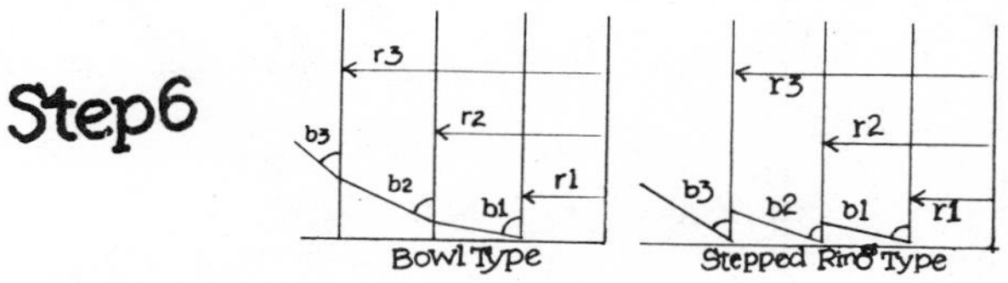

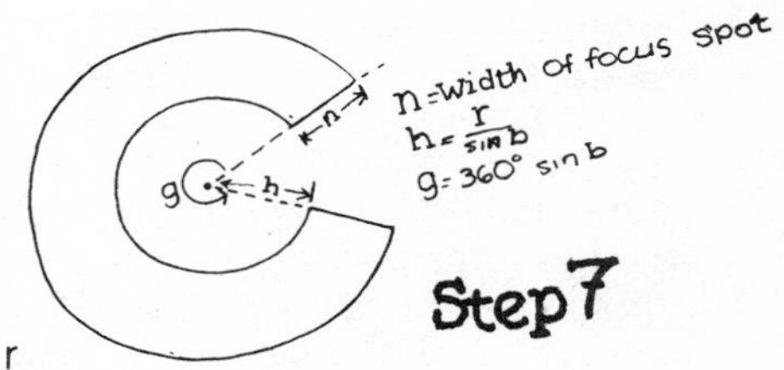

Step 6: Continue step 5 until the reflector is a wide as you want. If you are making a bowl-type reflector, each new ring will join onto the previous one. If you are making a stepped type reflector, each new ring will start from the baseline. Notice in the case of a stepped reflector, some of the rings may shade the others, especially if the reflector has a short focal length. One way to get around this is to leave a space between one ring and the next.

Now we need to know the radius and angle of the "C" shape that each ring will have when unrolled. Measure angle b and radius r for each ring.

n = width of focus spot

$$h = \frac{r}{\sin b}$$

$$g = 360° \sin b$$

The values of sin b can be easily found by looking them up in a table of trig functions. (You didn't think your teachers gave you trig just to be nasty, did you?)

Step 7: The rings can be cut from sheet aluminum, or for fair weather experiments, from cardboard and aluminum foil. Edmund Scientific sells large sheets of foil-backed cardboard as photographic reflectors. The rings can be assembled in any mechanically convenient fashion. The flat stepped configuration has an advantage here, since all the rings can be fastened to a single baseboard.

# A Prototype Solar Kitchen

C. J. SWET

*Released by the Professional Division of the American Society of Mechanical Engineers*

## ABSTRACT

A prototype solar kitchen is described that can provide high grade thermal energy for a variety of household uses. The heat is available indoors, in the evening, during periods of intermittent cloudiness, and in high winds, without manual positioning to follow the sun. The prototype unit is designed with the cooking and baking needs of a small family in mind, but the basic design is scalable to much larger heat delivery rates and adaptable to many other uses. Commonly available materials and components are used throughout, and no unusual skills are required for construction, installation, or use.

## INTRODUCTION

An earlier paper[1] presented conceptual designs for a universal solar kitchen, speaking to needs that are poorly met by existing solar cookers and ovens. The kitchen would provide high grade thermal energy for a variety of household uses. It could be used indoors, in the evening, in intermittent sunshine, and in high winds, without manual positioning to follow the sun.

A more recent paper[2] refined the technical approach, and plans are under way for the construction of a prototype unit intended to meet most of the cooking, baking, and ironing needs of a small family. The basic design is adaptable to many other uses and scalable to much higher heat delivery rates. It utilizes commercially available materials and standard components, and requires no unusual skills for construction, installation, or use.

This paper generally describes the prototype kitchen and discusses some special considerations. Its main objective is to show how advanced technology can be effectively applied to small scale decentralized energy production in ways that can be understood and implemented by the using community.

## GENERAL DESCRIPTION

Figs. 1 through 4 depict the overall configuration and characterizing features.

The line-focussing reflector concentrates incident solar energy onto the stationary heat pipe, about which it rotates to follow the sun. Sun tracking is fully automatic, being powered by direct solar radiation on the bimetal helix and controlled by the feedback sunshade.

The tilted heat pipe, which is aligned

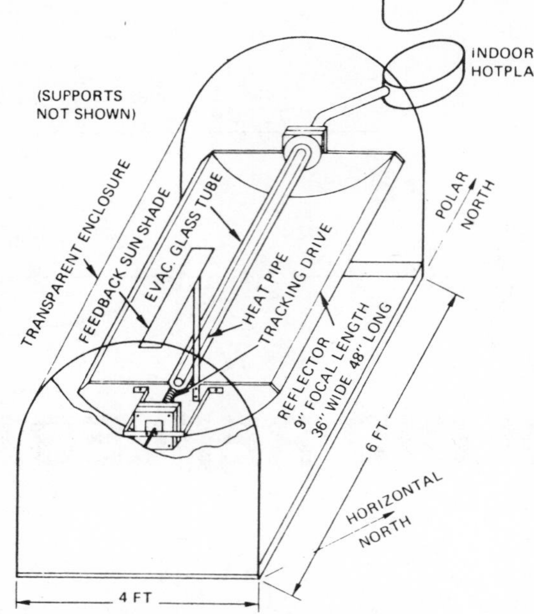


**Fig. 1** General Arrangement

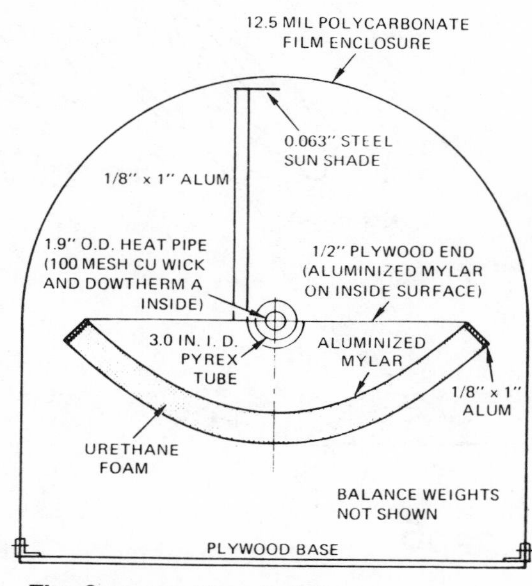


**Fig. 2** Section Through Collector

with the earth's rotational axis, provides near-isothermal heat transport at otherwise unachievable rates. It contains a small amount of Dow-therm A heat transfer fluid, which vaporizes in the irradiated section and condenses in the indoor hotplate. This fluid was chosen because of its moderate vapor pressure at the maximum expected operating pressure, and because it contracts when freezing without damage to the heat pipe. A valved connection is provided for filling and pressure measurement. Condensate returns by gravity and is distributed by wicking action over the entire inside surface. Rapid start-up in the morning is assured by the low fluid thermal inertia, and reverse heat flow is negligibly small when the sun is not in view.

The condensing hotplate can serve as a griddle, as a "burner" for cooking utensils, or as the heat source for a thermal storage mass which may be placed in an oven or used in the evening.

Radiative losses from the heat pipe and the bimetal helix are minimized by coating both components with surfaces of high solar absorptivity and low thermal emissivity. The "nickel-black" surfaces can be applied with conventional electroplating equipment. Convective losses are virtually eliminated by enclosing the heat pipe and helix in an evacuated glass tube. The required vacuum of about $10^{-4}$ torr can be produced by mechanical pumps. It will be increased, then maintained in the presence of small leakage, by calcium getter material. A valved connection is provided for evacuation and vacuum measurement. Rotary motion of the helix is transmitted to the collector by a synchronous magnetic drive, thus avoiding the need for moving seals.

The transparent polycarbonate enclosure minimizes tracking system torque requirements by shielding the collector from defocussing wind loads. It also protects the reflecting surface and the focal tube from damage and dust accumulation.

## SPECIAL CONSIDERATIONS

**Heat Balance.** When the hotplate is uncovered its surface temperature will ordinarily be hot enough for frying. Fig. 5 shows that it would be usefully hot over the expected range of griddle cleanliness, under conditions of moderate isolation.

When the thermal storage mass is on the hotplate and covered, the equilibrium condition will not create excessive pressures or exceed the recommended maximum operating temperature (750°F) for Dowtherm A, as shown in Fig. 5. The insulation is 2 inches of 85 percent magnesia.

Fig. 5 also indicates that heat delivery rates to cooking utensils will probably exceed 0.6 kw at water boiling temperatures and 0.4 kw at the fusion temperature of the thermal storage medium.

**Heat Storage** Two kinds of thermal storage capability are provided: to permit "coasting through" periods of intermittent cloudiness early in the day, and to furnish large quantities of heat for evening use.

The first kind of storage is provided by the thermal mass of the hotplate itself, which weighs about 20 lb. Under an insulating cover it can heat to nearly 500°F by about an hour after sunrise. As a bare griddle it would then take roughly 30 minutes to cool down to 300°F.

The second kind of thermal storage is provided by the removable sealed container of $LiNO_3$, which may be used either as a second cooking "burner" for later use or as the heat source for an oven. Roughly 4 hrs. of moderate insolation are required to heat it to and melt it at its fusion temperature of 490°F. It can then deliver 3200 Btu at constant temperature in an oven, or deliver 4680 Btu cooling to 300°F. Less expensive eutectic mixtures will also be considered.

**Sun Tracking.** The technology of thermal heliotropes is mature. Bimetal helices have been proposed as a means of orienting spacecraft solar arrays, and their use in that application has been verified by extensive analysis and model tests in a simulated orbital environment[3]. Their terrestrial use has also been investigated[4], and thermal response experiments have been conducted under atmospheric conditions[5].

The helix heats up whenever it absorbs incident solar energy faster than it loses heat to its surroundings. This causes unequal expansion of its two components and consequent rotation in the sun-following direction. Conductive losses can be made small and convection losses are near-zero in the vacuum environment. Radiation losses are minimized by the selectively absorbing outer surface.

The feedback shade moves with the collector to regulate helix insolation, which increases during the day to raise the equilibrium temperature and produce a mean angular rate matching that of the sun. On a clear day this tracking process continues until shortly before sunset, when full exposure to the reduced solar flux causes no further rise in temperature. At night the helix cools and returns to its pre-sunrise condition.

**Torque Transmission.** The helix must produce up to perhaps 1 in-lb. of torque to overcome bearing friction, mass imbalance, and collector inertia. To do so it must rotate sufficiently to lead the sun by the corresponding angular displacement of the magnetic drive. Fig. 6 is a representative daily history of temperature and torque production for the selected bimetal helix and magnetic drive. Note that the equilibrium temperature achievable with full insolation remains higher than that required for tracking until shortly before sunset.

**Tracking Errors.** Total enclosure eliminates the usual problems of defocussing wind loads, but the use of a thermal heliotrope introduces new sources of systematic tracking errors. These errors are determined by the helix diameter, the distance from the focal line to the feedback shade and the amount of helix shading required. This is illustrated in Fig. 7 for the representative conditions of Fig. 6. Note that the mean daily error is minimized by introducing a fixed bias in the sunshade alignment. During most of the day the error is less than the angle subtended by the solar disc, and less than 15 percent of that which corresponds to grazing incidence on the heat pipe. A shadow gauge will be mounted on the reflector to measure tracking accuracy.

**Sun Seeking.** Under conditions of intermittent cloudiness the helix cools and rotates in the reverse direction during each period of eclipse, then heats up and seeks the sun whenever it reappears. The cooling and reverse rotation are slow because of the low thermal emissivity; the catching up is relatively rapid because of the high solar absorptivity and the full insolation.

Sun seeking capability is mainly determined by the helix thermal response characteristics, which for a given bimetal are governed solely by its thickness and its surface properties. The time required to re-establish lock after a period of eclipse depends also on the degree and duration of eclipse, the time of day, the solar flux intensity, and conductive losses. This has not yet been studied in detail, but some feeling for potential performance is gained by examining the initial rotational rates that would be produced by momentary eclipse followed by full insolation. Fig. 8 is a daily history of these rates for the representative conditions of Fig. 6. For this configuration, the data suggest that recovery time would generally be less than a tenth of the time in eclipse.

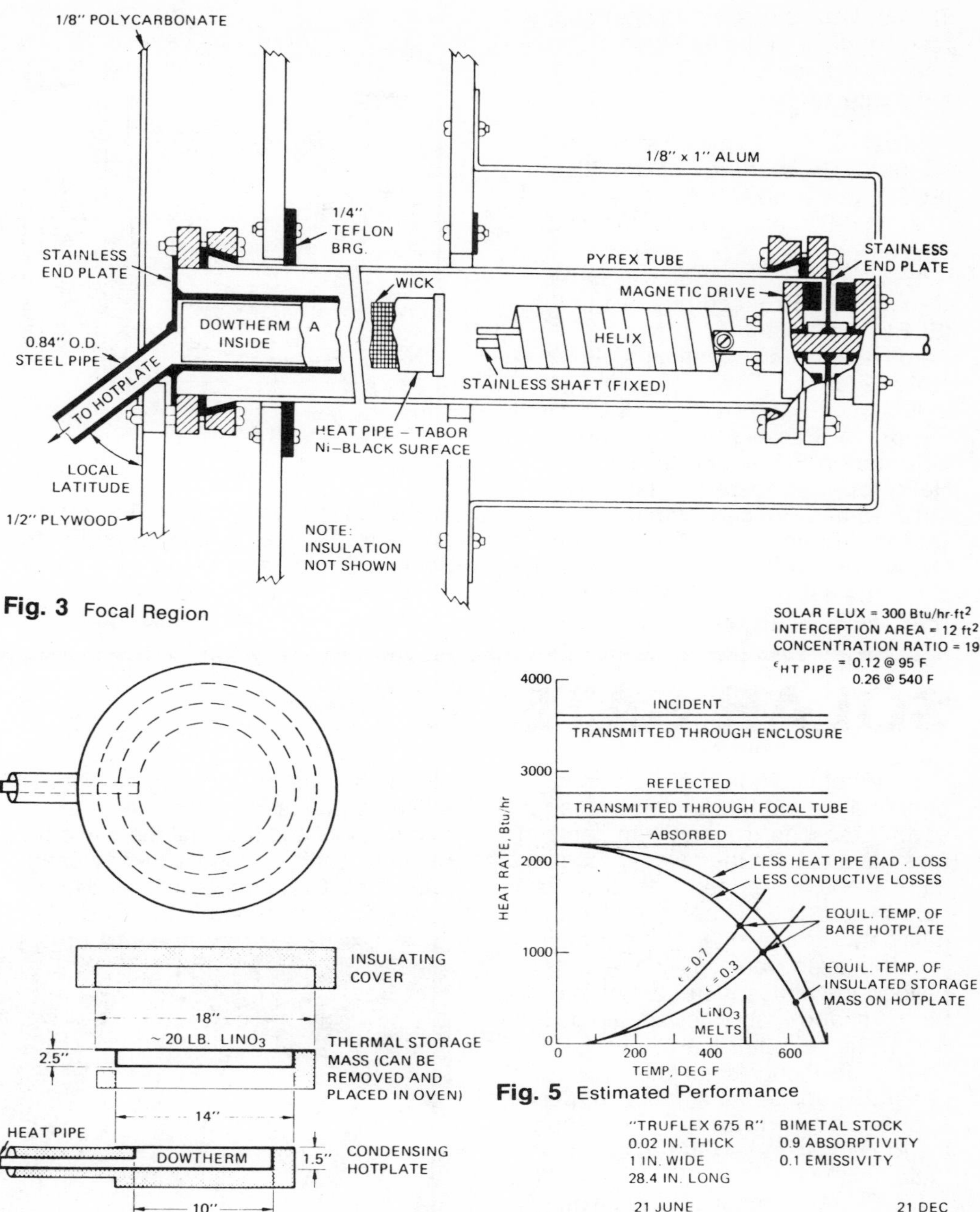


**Fig. 3** Focal Region

**Fig. 5** Estimated Performance

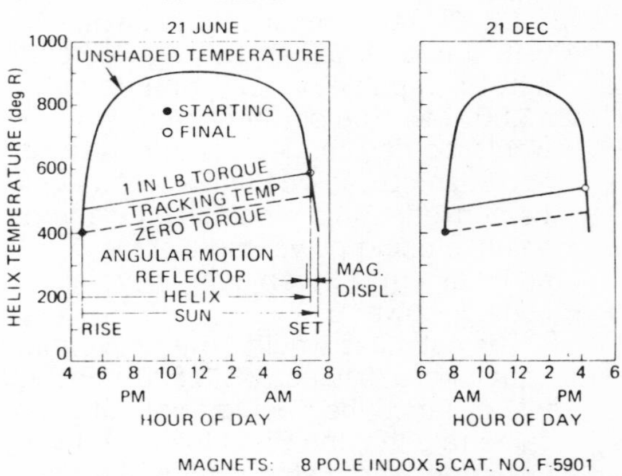


**Fig. 6** Tracking Limits

Thinner helix stock would shorten the recovery time at the expense of near-sunset torque availability.

## REFERENCES


[1]Swet, C. J., "A Universal Solar Kitchen", **Proceedings of the 7th IECEC**, paper no. 72912, American Chemical Society, Washington, D.C., 1972.

[2]Swet, C. J., "Heliotropic Thermal Generators", **Proceedings of the 8th IECEC**, American Institute of Aeronautics and Astronautics, New York, N.Y., 1973.

[3]Lott, D. R. and Byxbee, R. C., **Final Report for Passive Solar Array Orientation System (Thermal Heliotrope)**, Prepared by Lockheed Missiles and Space Company for Goddard Space Flight Center, Contract No. NAS 5-11637.

[4]Fairbanks, J. W. and Morse, F. H., "Passive Solar Array Orientation Devices for Terrestrial Application", Paper presented at International Solar Energy Society Conference, Greenbelt, Maryland, 10-14 May 1971.

[5]Morse, F. H., "Response Characteristics of a Thermal-heliotrope Solar-Array Orientation Device", **Proceedings of the 5th Aerospace Mechanisms Symposium**, Greenbelt, Maryland, 15-17 June 1970.


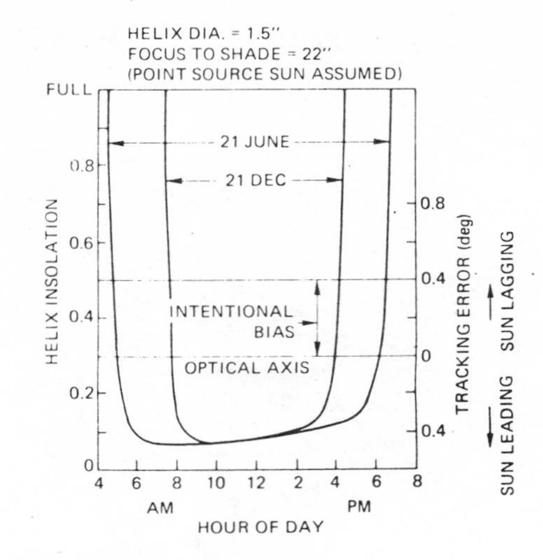


**Fig. 7** Tracking Errors

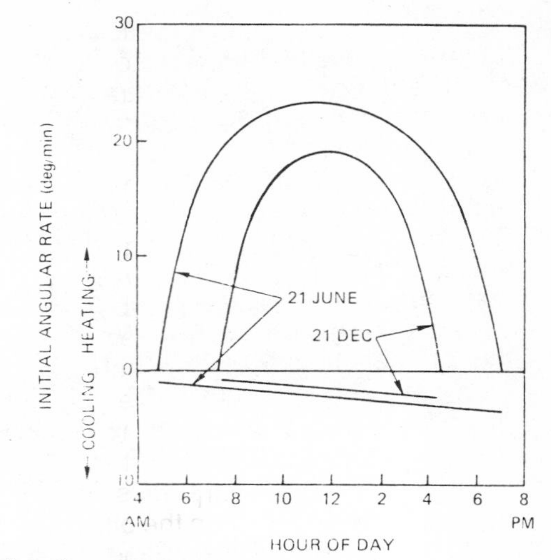


**Fig. 8** Thermal Response

# SOLAR SATELLITE

Solar cells turn in a disappointing performance at the Earth's surface for several reasons. To start with, the sun's rays have to fight their way through a thick atmosphere filled with dust, haze, clouds, etc. Then every night the Earth gets between the collectors and the Sun, which is even worse. Take the cells beyond the atmosphere, out of the Earth's shadow, and they should do from 6 to 15 times better than they would on Earth. With enough solar cells, you could generate substantial amounts of power and send it back to Earth. This, in short, is the idea of Dr. Peter Glaser, of Arthur D. Little, Inc.

Dr. Glaser proposes large satellites with solar cell panels twelve kilometers across, capable of delivering up to 15,000 Mw. The power captured in orbit would be transformed into microwaves and beamed back to Earth where it would be converted into electricity. One large satellite could power most of the northeastern U.S. replacing several nuclear power plants.

The satellites would orbit above the equator at a distance of 23,000 miles. At this distance they would rotate at the same rate that the Earth turns, so they would appear to stand still in the sky. They would be in almost perpetual sunlight. Only during a few nights around the time of the equinoxes would they pass into the Earth's shadow. An eclipse would last for slightly over an hour. Since it would take place in the middle of the night at a time of minimum power demand, it would cause little trouble.

The electrical power produced by the solar cells would be converted into microwaves by high efficiency microwave generators. An efficiency of 90% is hoped for. An antenna one kilometer across would focus the microwaves and transmit them in a tight beam to earth. At the frequency used (about 3.3 Ghz.) the beam would pass through the atmosphere, rainclouds and all, with almost no losses.

The beam would be aimed at a receiving station on the ground. Hundreds of thousands of small antennas would be connected together to form an array roughly seven kilometers

across. Built-in rectifying diodes would change the microwaves into direct current. Microwave power transmission has already been demonstrated in the laboratory.

Needless to say, the beam would have to be aimed very carefully. Automatic safeguards would be arranged to cut off the beam if it wandered away from the target area. The power distribution within the beam would be arranged so that it met safety standards for microwave exposure at its edges. Little is known about the effects of long term exposure to low intensity microwaves. It might be necessary to have airplanes detour around the beam, or on the other hand, it might be possible to use the land beneath the antennas for grazing.

A system of solar satellites would help to reduce our dependence on fossil fuel. By reducing the need for nuclear power plants it would help to relieve us of nagging fears about nuclear accidents. One unique advantage is a very low level of thermal pollution. The overall efficiency of the system is expected to be 70%, meaning that 30% of the total energy would be lost as waste heat. However most of this would take place in space. Only 10% of the waste heat would be released on the ground, and this would be spread over a large area. Present nuclear plants, by comparison, turn 60% of their energy into waste heat.

The deployment of a solar satellite would be a tremendous project. Nothing remotely like it has been done before. The satellite is hundreds of times larger and heavier than anything else ever sent into space. We are talking about hundreds of launchings and millions of tons in orbit. An improved version of the space shuttle would be needed to make the project possible. Each separate payload would have to be injected into the proper orbit, manuevered to the rendezvous point, deployed, and assembled. A lot will have to be learned before such a complex operation can be conducted successfully. Similarly, a great deal of work will have to be done to develop the improved components that will make the satellite possible. Low cost, light weight solar cells will have to be developed, and their manufacture will call a whole new industry into being.

The development of a solar satellite system would require a large initial investment. Prototype costs are expected to be comparable with those for the breeder reactor. After the development period, costs would be competitive with other power sources. Dr. Glaser expects that a satellite would return the energy expended in its construction in its first year of operation, after which it could be expected to have a useful life of 30 years.

The economic feasibility of the solar satellite depends to a great extent on future developments. Critics points out that Dr. Glaser's scheme is over-optimistic, in that it requires high efficiency, light weight, low cost components that do not yet exist. However none of the advances that Dr. Glaser forsees in 20 to 30 years are unreasonable if the preliminary development work needed is started now. It is likely that an early space shuttle mission will have experiments to test the feasibility of developing and transmitting power from space. Right now the solar satellite is only an interesting idea, but by the year 2000, it might be a useful energy alternative.


References

**Energy Research and Development and Space Technology**

Copies are available free while supplies last from:

Publication Clerk
Committee on Science and Astronautics
U.S. House of Representatives
Washington, D.C.


# SOLAR FARMS

Some views of the future include a solar collector on every rooftop. While the idea of self-sufficiency appeals to many people, there are drawbacks to this picture. It's an expensive way to do things. For everyone to buy, install, and maintain his own system would be wasteful duplication of effort. It is also inefficient. A large number of small systems will never be as efficient as one large system. Another difficulty is that few of today's homes can be adapted to solar energy.

Why not collect solar energy at a central location, convert it to electricity, and distribute it where needed? A large central plant could be maintained and operated much more efficiently than numerous small plants. The plant could be easily integrated into our present electrical distribution system. This is the idea of Dr. Aden Meinel, director of the Optical Science Center at the University of Arizona. Dr. Meinel proposes that solar energy farms be established in desert areas of the U.S. The farms would consist of large areas of solar collectors that would heat a fluid such as liquid sodium to a temperature of about 200°C (400°F). The hot fluid would be used to generate electricity in conventional fashion. Tanks of eutectic salts would store heat for use at night and on cloudy days.

Such a system would produce no pollution or radiation hazard. It might be a problem to cool its condensers properly and dispose of the waste heat. The plant would not be likely to have streams available to it for cooling water. Dr. Meinel suggests using the waste heat for such tasks as desalting water. He points out that it is possible to engineer the system so that the thermal balance of the area remains unaffected.

One drawback to the proposal is that it requires a lot of land. A one megawatt plant would require roughly ten square miles. However, the area required to fill the energy needs of the U.S. in this way is small when compared with the land scarred by strip mining, and tiny when compared with the amount of land used for farming. Dr. Meinel expects costs to be comparable with present methods of producing electricity, though he admits that maintenance costs would probably be somewhat higher, since a solar farm would have a field of collectors to maintain plus all the machinery of a conventional power plant.

In Dr. Meinel's original proposal, the collectors consisted of long tubes with selective coatings. Mirrors or lenses helped the collectors reach the high temperatures needed for efficient operation. The higher the operating temperature, the greater the overall efficiency of the system. Dr. Meinel found, however, that the effects of clouds and haze upon a test collector were greater than he had anticipated. Recent improvements in selective surfaces have given him the hope that the necessary temperature can be attained through the use of selective surfaces alone. Doing away with the focusing elements and the machinery needed to keep the collectors pointed at the sun would make the system cheaper and more reliable. Dr. Meinel is now experimenting with a new type of collector which uses reflectors not to focus the sun's rays, but to make the collector relatively independent of the sun's direction.

Dr. Meinel has proposed that a one acre demonstration system be built. Such a system would answer questions about total energy yield, reliability, and economic feasibility. If the prototype were successful, solar energy farms could become part of our lives in a relatively short time.

### Selected Bibliography

A lot of books on energy are coming out. The reader must be on his guard for bright-eyed daydreams posing as practical solutions. Many of the most interesting projects are still in the demonstration stage. We've tried to select books that are interesting, useful, and currently available. Most of them have bibliographies of their own.

#### Periodicals

Alternative Sources of Energy
A.S.E.
Rt. 2, Box 90A
Milaca, Minnesota 56353
$5.00/6 issues

*Alternative Sources of Energy is a newsletter for people concerned with the development of alternative technologies. It was the first publication of its kind. Its main purpose is to serve as a communications network for the exchange of ideas and information. Readers are encouraged to send in articles and letters; most of the information in A.S.E. comes from readers.*

Undercurrents
Undercurrents Ltd.
275 Finchley Road
London NW3 England
$4.00/6 issues (surface mail)

*This is the British opposite number of A.S.E. It's more diverse and more polished, takes itself less seriously, and contains fewer do-it-yourself details.*

#### General

Alternative Sources of Energy — Book One-practical technology and philosophy for a decentralized society.
Edited by Sandy Eccli and others

Eugene and Sandy Eccli
928 Second Street S.W. Apt. 4
Roanoke, Virginia 24016
$4.00 - 60 pp.

*Here are all the articles from the first ten issues of A.S.E., rearranged into categories, expanded with new material, and printed in a larger and clearer format. It makes an impressive pile of information. This is unquestionably the best all-round book on the subject, both for the builder-experimenter as well as the interested reader.*

The Mother Earth News Handbook of Homemade Power
by the Mother Earth News Staff

The Mother Earth News
P.O. Box 70
Hendersonville, North Carolina 28739
$1.95 - 374 pp.

*This book is a collection of articles from the Mother Earth News. There are chapters on wood, water, wind, sun, and methane. It will be most useful to the homesteader, but others will find it interesting as well. Most of the material is introductory, but there are a few interesting how-to's also.*

The Natural Energy Workbook
by Peter Clark

Village Design
1545 Dwight
Berkeley, Calif. 94703
$2.95 - 44 pp.

*This is basically an idea book. It introduces a wide variety of energy alternatives and discusses them briefly. Practical details are few. An interesting exception is a section which describes the making of low cost parabolic reflectors for solar energy experiments. The appearance of the book is unfortunately marred by a poor printing job.*

Survival Scrapbook #3 - Energy
by Stefan A. Szczelkun

Schocken Books
200 Madison Avenue
New York, N.Y. 10016
$3.95 - 57 pp.

*This is another idea book. It ranges more widely than most, getting into esoteric topics like mental energy. No one topic is covered in much detail.*

Exploring Energy Choices
A Preliminary Report of the Ford Foundation's Energy Policy Project

The Energy Policy Project
P.O. Box 23212
Washington, D.C. 20024
$.75 - 81 pp.

*This report gives an excellent summary of our present energy situation. It is well documented, with plenty of facts and figures. The most interesting part, however, is an examination of three possible energy futures which could result from today's decisions. The first, which assumes that present trends continue, strains our resources to the utmost. The second uses conservation and high efficiency technology to reduce demand. In the third, energy growth is reduced to zero through changes in our way of life.*

Engineer's Dreams
by Willy Ley

Viking Press
625 Madison Avenue
New York, N.Y. 10022
$4.50 - 280 pp.

*Engineers dream on a grand scale and their concepts frequently go beyond what is presently possible in our mundane world. About half this book is devoted to energy. While this book is somewhat dated now, it's exciting to see how many of these once far-out ideas are becoming reality.*

Energistics
by W. J. Cox

W. J. Cox
Policy Research Group
Public Works
Ottawa, Ontario,
Canada
free - 154 pp.

*In this highly interesting book, philosophy is applied to questions of energy use, conservation and sensible design. It's full of delightful quotes, insights, and speculations.*

#### Solar

Direct Use of the Sun's Energy
Farrington Daniels

Ballantine Books
P.O. Box 505
Westminster, Maryland 21157
$2.20 - 271 pp.

*Though first published in 1964, this is still the best book on solar energy. Daniels gives clear summaries of the work that has been done in every part of the field. The book is detailed, but not overly technical.*

Solar Energy as a National Energy Resource
(Stock No. 3800-00164)
NSF/NASA Solar Energy Panel

Superintendent of Documents
U.S. Government Printing Office
Washington, D.C. 20402
$1.20 - 85 pp.

*This book is an extensive discussion of all the applications solar energy may find in the future.*

Solar Heated Building - A Brief Survey
by William A. Shurcliff

Solar Energy Digest
P.O. Box 17776
San Diego, Calif. 92117
$4.00 - 36 pp.

*Mr. Shurcliff is doing his best to make a complete survey of structures that are fully or partially solar heated. Each entry consists of a sketch of the structure and a brief description of its operating principles. Where available, details of cost, performance, troubles encountered, and present status are given, along with references. The survey presently includes 58 buildings, but it is presently being expanded to include half a dozen more.*

Proceedings of the Solar Heating and Cooling for Buildings Workshop
no. PB-223-536

National Technical Information Service
U.S. Dept. of Commerce
Springfield, Virginia 22151
$3.00 - 226 pp.

*This book is an interesting miscellany of reports on various aspects of solar heating and cooling. None of them are very long and most are fairly technical.*

**Wind**

Wind & Windspinners
Michael A. Hackleman

Earthmind
Josel Drive
Saugus, Calif. 91350
$7.50 - 115 pp.

*This book is an essential one for anyone interested in home built wind generators. It covers such subjects as alternator basics, the care and charging of batteries, the testing of used batteries, regulators and control systems, wiring, etc. Practical details on the construction of a Savonius rotor are provided, but the information in this book also applies to other types of wind generators. The style is informal, but the information is clear and complete and inspires confidence in the reader.*

Wind Energy Conversion Systems
NSF/RA/W-73-006
Edited by Joseph M. Savino

NASA - Lewis Research Center
21000 Brook park Road
Cleveland, Ohio 44135
Free - 258 pp.

*In December 1973 a joint NSF/NASA workshop was held on wind energy. Proponents of alternate technologies rubbed shoulders with engineers. The resulting proceedings show current thinking about wind energy from a wide variety of viewpoints.*

**Methane**

Methane Digesters
for fuel gas and fertilizer

The New Alchemy Institute

The New Alchemy Institute - East
Box 432
Woods Hole, Mass. 02543

The New Alchemy Institute - West
15 W. Anapamu
Santa Barbara, Calif. 93101
$3.00 - 47 pp.

*This is the best book around on methane generators. It covers the whole process of digestion in detail. Two simple small digesters for learning on are described, one using oil drums and one a truck inner tube.*

**Bio-Gas Plant**

Generating Methane From Organic Wastes
Ram Bux Singh

James Whitehurst
Benson, Vermont 05731
$5.00 - 70 pp.

*India has been investigating methane as a supplement for scarce fuel supplies. This book contains recommendations for generating methane on a practical basis based on experience from many gas plants. Most of the data concerns very large digesters.*

A Homesite Power Unit: Methane Generator
Les Auerbach

Les Auerbach
242 Copse Road
Madison, Conn. 06443
$5.00 - 50 pp.

*This book gives a short description of the digesting process and equations for determinging the size of a digester and its storage tank. Also given are the details of the construction and operation of a digester made from used components.*

**Batteries**

Automobile Batteries - Their Selection and Care
Stock No. 220-0067

Superintendent of Documents
U.S. Government Printing Office
Washington, D.C. 20407
$.40 - 13 pp.

*A simple explanation of how the auto battery works and how to care for it.*

The Storage Battery Lead-Acid Types
The Nickel-Cadium Storage Battery
Exide Power Systems Division

Exide Power Systems Div.
E S B Inc.
Philadelphia, Pa. 19120

| | | |
|---|---|---|
| Lead Acid | 33 pp. | $1.00 |
| Ni-Cad | 23 pp. | $1.00 |

*These two booklets give detailed description of how storage batteries work, what to expect from them, and how to care for them.*

**Water**

Cloudburst
A Handbook of Rural Skills & Technology
Edited by Vic Marks

Cloudburst Press
Box 79
Brackendale, B.D.
Canada
$3.95 - 126 pp.

*Although this book is not primarily about energy, it has a very good section on the design and construction of small water wheels, as well as other items of interest to the do-it-yourself energy enthusiast.*

U.S. Government

A free price list of energy related publications is available from the Superintendent of Documents. Ask for "Selected Publications Relating to Energy Conservation."

The Superintendent, however, handles only a small part of the documents published by the government. When energy experts testify before Congress, for instance, the results are printed in booklet form for later reference by the Congressmen. These booklets are also available to the public, free for the asking, as long as the supply holds out. Titles can be looked up in the Monthly Index to U.S. Government Publications, which is carried by most public libraries. Address your request to the Publications Clerk of the Committee concerned.

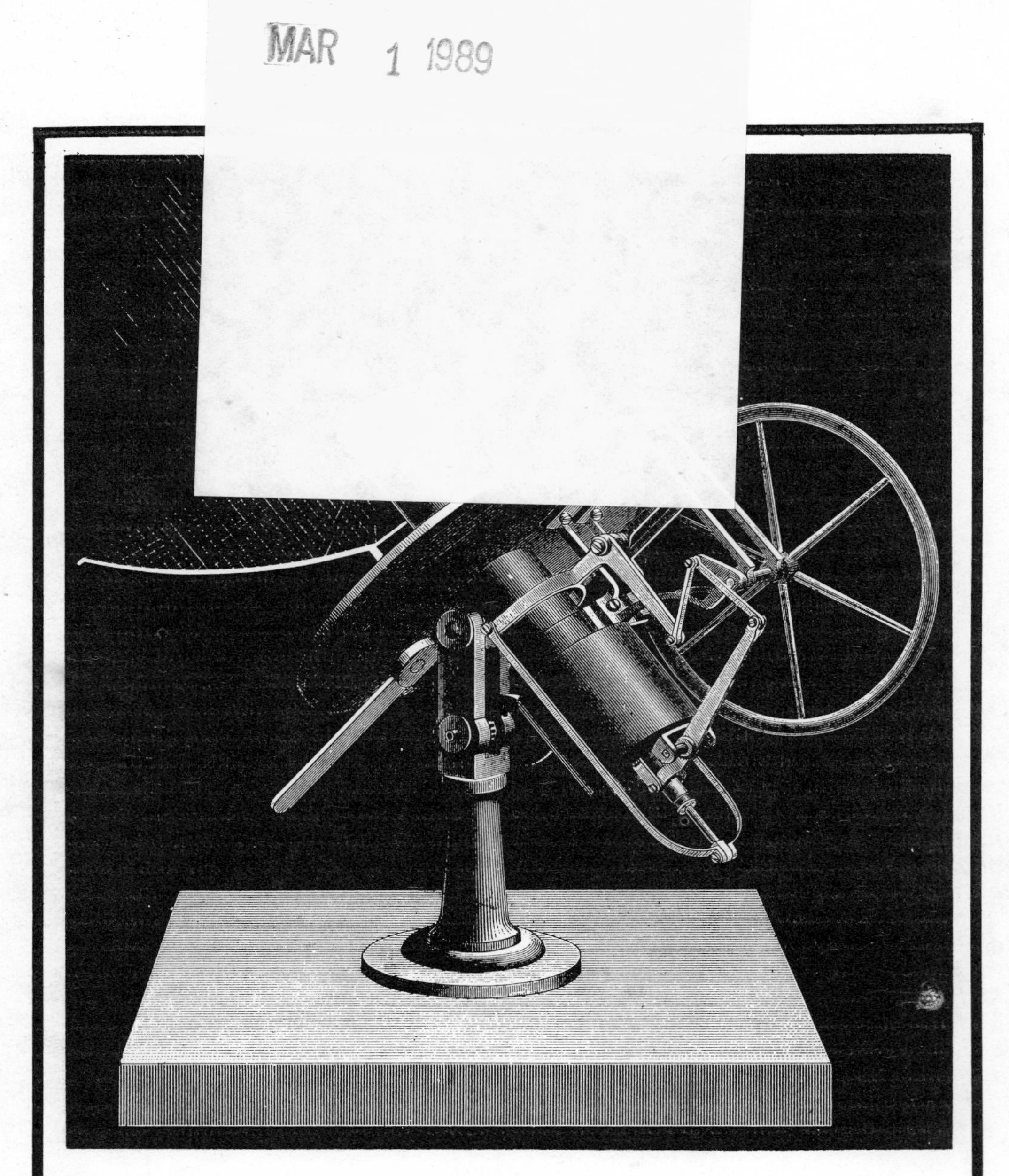

**Last Word**

We hope that you have enjoyed this book. We'd appreciate your comments and suggestions. Some of you may wonder why we didn't include such-and-such. It's probably because we either couldn't get any good material on the subject or because we didn't hear about it in time. It's difficult to keep up with all that happens in the energy field, especially since many amateur experiments never get much publicity. You can help us by bringing news of interesting new energy developments to our attention. Even negative information about a project that flops is useful if it keeps others from making the same mistake. We've called this **Energy Book #1** because we hope to follow it with a bigger and better **Energy Book #2**.

*John Prenis*
*161 W. Penn St.*
*Phila., Pa. 19144*

*Stuart Teacher*
*Running Press*
*38 South 19th St.*
*Phila., Pa. 19103*